PRINCETON AERONAUTICAL PAPERBACKS

PRINCETON UNIVERSITY PRESS · PRINCETON, N.J.

NUMBER 8
PRINCETON AERONAUTICAL
PAPERBACKS
COLEMAN duP. DONALDSON, GENERAL EDITOR

———————◆————————

FLOW OF
RAREFIED GASES

———————◆————————

BY SAMUEL A. SCHAAF
AND PAUL L. CHAMBRÉ

PRINCETON, NEW JERSEY
PRINCETON UNIVERSITY PRESS
1961

PRINTED IN THE UNITED STATES OF AMERICA

HIGH SPEED AERODYNAMICS

AND JET PROPULSION

———— •◆• ————

BOARD OF EDITORS

THEODORE VON KÁRMÁN, *Chairman*
HUGH L. DRYDEN
HUGH S. TAYLOR

COLEMAN DUP. DONALDSON, General Editor, 1956–
Associate Editor, 1955–1956

JOSEPH V. CHARYK, General Editor, 1952–
Associate Editor, 1949–1952

MARTIN SUMMERFIELD, General Editor, 1949–1952

RICHARD S. SNEDEKER, Associate Editor, 1955–

PRINCETON, NEW JERSEY
PRINCETON UNIVERSITY PRESS

PREFACE

The favorable response of many engineers and scientists throughout the world to those volumes of the Princeton Series on High Speed Aerodynamics and Jet Propulsion that have already been published has been most gratifying to those of us who have labored to accomplish its completion. As must happen in gathering together a large number of separate contributions from many authors, the general editor's task is brightened occasionally by the receipt of a particularly outstanding manuscript. The receipt of such a manuscript for inclusion in the Princeton Series was always an event which, while extremely gratifying to the editors in one respect, was nevertheless, in certain particular cases, a cause of some concern. In the case of some outstanding manuscripts, namely those which seemed to form a complete and self-sufficient entity within themselves, it seemed a shame to restrict their distribution by their inclusion in one of the large and hence expensive volumes of the Princeton Series.

In the last year or so, both Princeton University Press, as publishers of the Princeton Series, and I, as General Editor, have received many enquiries from persons engaged in research and from professors at some of our leading universities concerning the possibility of making available at paperback prices certain portions of the original series. Among those who actively campaigned for a wider distribution of certain portions of the Princeton Series, special mention should be made of Professor Irving Glassman of Princeton University, who made a number of helpful suggestions concerning those portions of the Series which might be of use to students were the material available at a lower price.

In answer to this demand for a wider distribution of certain portions of the Princeton Series, and because it was felt desirable to introduce the Series to a wider audience, the present Princeton Aeronautical Paperbacks series has been launched. This series will make available in small paper-backed volumes those portions of the larger Princeton Series which it is felt will be most useful to both students and research engineers. It should be pointed out that these paperbacks constitute but a very small part of the original series, the first seven published volumes of which have averaged more than 750 pages per volume.

For the sake of economy, these small books have been prepared by direct reproduction of the text from the original Princeton Series, and no attempt has been made to provide introductory material or to eliminate cross references to other portions of the original volumes. It is hoped that these editorial omissions will be more than offset by the utility and quality of the individual contributions themselves.

Coleman duP. Donaldson, General Editor

PUBLISHER's NOTE: Other articles from later volumes of the clothbound series, *High Speed Aerodynamics and Jet Propulsion*, may be issued in similar paperback form upon completion of the original series.

CONTENTS

SECTION H

FLOW OF RAREFIED GASES

S. A. SCHAAF

P. L. CHAMBRÉ

CHAPTER 1. INTRODUCTION

H,1. Introduction. The mechanics of rarefied gases has been the subject of many investigations since the time of Maxwell [*1*, pp. 26, 681]. Until comparatively recently, these studies were confined to the case of very slow speeds and in general to "internal" flow geometries associated typically with vacuum installations. Summaries are given in I,I and in standard texts on the kinetic theory of gases, e.g. Kennard [*2*] or Loeb [*3*]. Since about 1946, there has been a considerable revival of interest and activity in the field, largely due to the possibility of flight at very high altitudes and at very high speeds. A growing body of both experimental and theoretical results has been obtained, with emphasis on the aerodynamic problem of very high speed flow of a rarefied gas past a body which is wholly "submerged." This activity was given impetus in pioneering articles by Tsien [*4,5*], which formulated many of the significant problems as well as the general lines of research which have subsequently proved successful.

A rarefied gas flow is a flow in which the length of the molecular mean free path $\bar{l}$ is comparable to some significant dimension L of the flow field. The gas then does not behave entirely as a continuous fluid but rather exhibits some characteristics of its coarse molecular structure. The dimensionless ratio $\bar{l}/L$ is denoted by K and is called the Knudsen number. A rarefied gas flow is thus one for which K is not negligibly small. For some considerations, L may be a characteristic dimension of the body itself or the diameter of an internal flow conduit. For other considerations, L may be the boundary layer thickness, the diameter of a wind tunnel probe, or the thickness of the shock transition zone. In particular, one may expect to encounter rarefied gas effects in those regions of the flow possessing very sharp gradients, i.e. regions in which the velocity, pressure, or temperature change appreciably in the space of a few mean free paths regardless of whether or not the absolute density of the gas flow is especially low.

The Knudsen number K is related to the more familiar parameters of fluid mechanics, the Mach number M and the Reynolds number Re. From kinetic theory, one defines $\bar{l}$ by the relation (Eq. 1-1)[1]

$$\nu = \tfrac{1}{2}\bar{l}\bar{v}_m \tag{1-1}$$

where ν is the kinematic viscosity and $\bar{v}_m$ is the mean molecular speed. The mean speed $\bar{v}_m$ is related to the sound speed a as follows:

$$a = \bar{v}_m \sqrt{\frac{\pi\gamma}{8}} \tag{1-2}$$

where γ is the isentropic exponent; hence one obtains

$$\bar{l} = 1.26 \sqrt{\gamma}\,\frac{\nu}{a} \tag{1-3}$$

Combining these results yields the fundamental relation

$$K = 1.26 \sqrt{\gamma}\,\frac{M}{Re} \tag{1-4}$$

where K and Re are both based on the same characteristic length L.

H,2. Flow Regimes. The division of gas dynamics into various regimes, based on characteristic ranges of values of an appropriate Knudsen number, has been proposed by Tsien [4], Roberts [6], Donaldson [7], and Siegel [8], among others. The various criteria which have been suggested are in some disagreement. In the present treatment the terms "continuum flow," "slip flow," "transition flow," and "free molecule flow" are defined so that they correspond to flows in which, roughly speaking, the density levels are, respectively, ordinary, slightly rarefied, moderately rarefied, and highly rarefied.

For flows of high Reynolds number, i.e. $Re \gg 1$, the significant characteristic dimension of the flow field, which is of importance in the determination of viscous effects, is the boundary layer thickness δ rather than a dimension L typical of the body itself. Since

$$\frac{\delta}{L} \sim \frac{1}{\sqrt{Re}} \tag{2-1}$$

the corresponding Knudsen number K is given by

$$K \sim \frac{M}{\sqrt{Re}} \tag{2-2}$$

Ordinary gas dynamics hence prevails for $M/\sqrt{Re} \ll 1$ and $Re \gg 1$. On

[1] The factor $\tfrac{1}{2}$ is sometimes replaced by 0.499; the present application does not seem to warrant such a refinement.

the other hand, for very small Reynolds numbers, the Stokes type "slow flow" occurs and the characteristic dimension itself is the significant parameter. Also, for any internal flow, only the diameter L of the conduit is of significance. Here the appropriate Knudsen number is simply K based on the body dimension and ordinary low speed continuum dynamics prevails for $M/Re \ll 1$. For flows in which the value of the appropriate Knudsen number is small but not negligible, some departure from

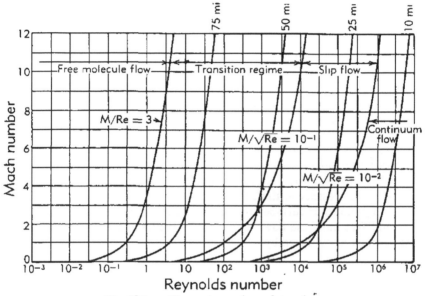

Fig. H,2a. The regimes of gas dynamics.

continuum gas dynamics phenomena may be expected to occur. As shown in more detail below, one of the more striking of these effects is the phenomenon of "slip"; the layer of gas immediately adjacent to a solid surface is no longer at rest but has a finite tangential velocity. The term "slip flow" is thus appropriate for flows of small but not negligible Knudsen number. The change from ordinary continuum gas dynamics to this regime is of course gradual; but to fix ideas, the slip flow regime on the basis of our present experimental evidence is defined by the following limits:

$$0.01 < \frac{M}{\sqrt{Re}} < 0.1, \quad Re > 1$$

$$0.01 < \frac{M}{Re} < 0.1, \quad Re < 1$$

(2-3)

These boundaries are depicted in Fig. H,2a.

In the slip flow regime, so defined, the mean free path is of the order of 1 to 10 per cent of the boundary layer thickness or other characteristic dimension of the flow field. Slip flow effects may thus be expected to be approximately of this same order. A typical example, for the slip correction in the slow speed flow, can be seen in Eq. 17-3. For the boundary layer range, consult Eq. 19-5. It may be observed that both these examples are in agreement with the above criterion, Eq. 2-3. It should be noted that in the slip flow regime, either the Reynolds number must be very small or the Mach number large. Thus true rarefaction effects such as slip occur only in coincidence with either strong viscous or compressibility effects. In the slip flow regime, as defined above, these latter phenomena very often dominate rarefaction effects associated with the coarse molecular structure of the gas, and really large scale deviations from continuum behavior are then not apparent until the "transition" regime, defined below, is reached. It might also be observed that in the hypersonic range, the boundary layer thickness is no longer given by Eq. 2-1, so that appropriate modifications in the parameter $M/\sqrt{Re}$ should be made here.

For extremely rarefied flows, the mean free path $\bar{l}$ is much greater than a characteristic body dimension L. Under these circumstances no boundary layer is formed. Molecules reemitted from a surface do not collide with free stream molecules until far away from the body. One may consequently neglect any distortion of the free stream velocity distribution due to the presence of the body. Here the flow phenomena are mostly governed by molecule-surface interaction. This regime of fluid mechanics is termed "free molecule flow" and may be defined on the basis of present experimental evidence by

$$\frac{M}{Re} > 3 \tag{2-4}$$

In the transition regime between the slip flow and free molecule regimes, the mean free path is of the same order as a typical body dimension. Surface collisions and free stream intermolecular collisions are of more or less equal importance, and the analysis becomes extremely complicated. Present knowledge about this transition regime is very much more limited than that in the free molecule or the slip regime. Except for a few special theoretical results which are presented in subsequent articles, the information which is available is mostly empirical. For some theoretical considerations, that portion of the transition regime immediately adjacent to the free molecule flow regime becomes of special importance. It corresponds physically to the region in which the effects of a few intermolecular collisions begin to distort the free stream velocity distribution. This regime has not been given a separate designation, however, and in the present section it is not considered separately. To summarize: the terms "slip flow," "transition flow," and "free molecule flow" as defined above are considered sufficient to designate flows of slight, moderate, and great rarefaction, respectively.

These various flow regimes are shown in Fig. H,2a, together with the corresponding altitudes in miles above sea level, provided that the characteristic body dimension is taken as 1 foot. Present knowledge as to the constitution and density of the upper atmosphere, and consequently of the variation of the mean free path with altitude, is somewhat limited [9,10]. The calculations are based on the NACA "proposed" standard atmosphere tables [10]. The results are given in Fig. H,2b.

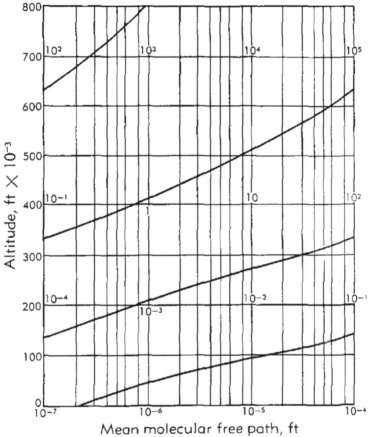

Fig. H,2b. Mean molecular free path of the adopted atmosphere.

A few numerical examples drawn from Fig. H,2a and H,2b may serve to relate these phenomena to typical conditions. Using 10 feet as a characteristic dimension for a high altitude missile, slip effects are expected to begin to be of importance at an altitude of 30 miles for $M = 1$ and at 20 miles for $M = 4$. Free molecule flow is established at about 90 miles for any Mach number. Compressible laminar boundary layer studies with a free stream Mach number greater than about 3 may be expected to be accompanied by some slip flow effects if the Reynolds number is of the order of 10^5 or less. Likewise, hypersonic wind tunnel design, operation, and instrumentation may involve slip flow phenomena unless very high stagnation pressures are used. For example, if the stagnation pressure is 100 atmospheres at normal temperature, the Reynolds number per inch in a test section at $M = 10$ is of the order 300,000. This is well within the slip regime. Very small hot wire probes used in boundary layer measurements or in turbulence studies may exhibit rarefaction effects at wind tunnel densities only slightly less than normal [11]. It might also be appropriate to recall that the mean free path is not the distance between molecules. At an altitude of 80 miles, for example, the mean free path is about one foot, but there are still 10^{13} molecules per cm^3. Most meteors are entirely vaporized by air friction at altitudes of this order, entirely in free molecule flow.

CHAPTER 2. FREE MOLECULE FLOW

H,3. Introduction. The free molecule flow regime is the regime of extreme rarefaction. The molecular mean free path is by definition many times the characteristic dimension of the body which is assumed to be located in a gas flow of infinite extent. The molecules which hit the surface of the body and are then reemitted on the average travel very far before colliding with other molecules. It is consequently valid to neglect the effect of the reemitted particles on the incident stream, at least so far as effects on the body itself are concerned. The incident flow is therefore assumed to be entirely undisturbed by the presence of the body. This is the basic assumption of free molecule flow theory. It is a consequence of this basic assumption that no shock waves are expected to form in the vicinity of the object. The "boundary layer" will be very diffuse and has no effect on the flow incident on the body.

Theoretical calculations of the external heat transfer and aerodynamic characteristics of bodies submerged in a free molecule flow field may be carried out by treating the flows of incident and reflected or reemitted molecules separately. In calculating the flux of momentum or energy incident on the surface it is assumed that the approaching gas is in local Maxwellian equilibrium. The results should hence be applied to very high altitude considerations with some care, since present knowledge as to the state and composition of the upper atmosphere is limited.

Early external free molecule flow calculations by Epstein [12] were confined to the case of small mean flow velocities while those of Zahm [13] treated the other extreme, neglecting the thermal motions of the molecules as compared to their mean flow. There is a close connection between this limiting case of free molecule flow and the "Newtonian flow" approximation of hypersonic theory [14]. More recently, calculations which take proper account of both the thermal and macroscopic motions have been carried out by a number of investigators. The treatment presented in this chapter follows, to a considerable extent, that of Stalder, et al. [15,16], Tsien [4], and Sänger [17].

H,4. Reflected Molecules. The determination of the flux of momentum or energy carried by molecules reflected or reemitted from the body surface requires a specification of the interaction between the impinging particles and the surface. The complete specification of this interaction would necessitate the theoretical or experimental determination of the velocity distribution function of the reemitted molecules, given their incident velocities. Fortunately, it is not necessary to have such detailed information in the free molecule flow regime. It is sufficient to know only certain average parameters which characterize the interaction phenomena and which are known as the accommodation and reflection coefficients.

For energy considerations one utilizes the "thermal accommodation coefficient" α, introduced by Smoluchowski [18] and Knudsen [19], which is defined by

$$\alpha = \frac{dE_i - dE_r}{dE_i - dE_w} \tag{4-1}$$

dE_i and dE_r are respectively the energy fluxes incident on and reemitted from a differential surface element per unit time. The quantity dE_w is the energy flux that would be carried away if all incident molecules were reemitted with a Maxwellian distribution corresponding to the surface temperature T_w. The term α is thus a measure of the degree to which the molecules have their mean energy "accommodated" to what it would be if the returning molecules were issuing with energy dE_w. For perfect accommodation $\alpha = 1$, while for the case of vanishing energy exchange with the surface, $dE_i = dE_r$ and hence $\alpha = 0$. It is implicitly assumed in the definition of α that all the energies associated with those molecular degrees of freedom which enter into an energy exchange with the surface are accommodated to the same degree. Available experimental evidence seems to indicate that this is approximately true for the translational and rotational energy components, while the vibrational energy is affected to a much less extent in a surface collision (see Herzfeld [20, pp. 228–251]). If necessary, it is possible to introduce separate accommodation coefficients for each energy component. Such refinements are, however, not considered in the present treatment. Experimentally determined values of α for various typical air-surface combinations as obtained by Wiedmann [21] are given in Table H,4a. See also Devienne [22,23].

The traditional treatment of momentum transfer originally due to Maxwell [1, p. 708] and subsequently followed in many free molecule flow calculations will be slightly modified. It has been customary to suppose that a fraction $(1 - \sigma)$ of the incident molecules is reflected "specularly." This reflection consists of a reversal of the normal velocity component which produces a normal momentum transfer to the surface and leaves the tangential velocity component unchanged. The remaining incident fraction σ is assumed to be reflected "diffusely," i.e. the molecules issue with a Maxwellian velocity distribution at a temperature not necessarily equal to that of the surface. It is this fraction σ which contributes the tangential momentum transfer to the surface.

Table H,4a. Thermal accommodation coefficient α for air [21].

Surface	α
Flat lacquer on bronze	0.88–0 89
Polished bronze	0.91–0.94
Machined bronze	0.89–0 93
Etched bronze	0.93–0.95
Polished cast iron	0.87–0 93
Machined cast iron	0 87–0 88
Etched cast iron	0 89–0 96
Polished aluminum	0 87–0 95
Machined aluminum	0.95–0.97
Etched aluminum	0 89–0.97

There is, however, preliminary experimental evidence that the interaction for air on typical engineering surfaces is not of such a simple type [24]. Consequently the single parameter σ may not suffice to describe the reflection phenomenon completely. In order to specify both the tangential and normal force components of the reflected flux the following two coefficients are introduced in analogy to Eq. 4-1

$$\sigma = \frac{\tau_i - \tau_r}{\tau_i - \tau_w}, \quad (\tau_w = 0) \tag{4-2}$$

$$\sigma' = \frac{p_i - p_r}{p_i - p_w} \tag{4-3}$$

where τ and p are respectively the tangential and normal momentum components. The subscripts i and r refer to the incident and reflected flux while p_w and τ_w denote respectively the normal and tangential momentum components of the molecules which are reemitted with a Maxwellian distribution at the surface temperature T_w. (Obviously τ_w must be zero.)

For the hypothetical cases of (i), entirely specular reflection with vanishing energy exchange, one would have $\alpha = \sigma = \sigma' = 0$, while for (ii), entirely diffuse reflection, which has been completely accommodated to the surface temperature T_w, one would have $\alpha = \sigma = \sigma' = 1$. In general, it is to be expected that the three parameters will be independent.

No information is available at the present time with respect to the quantity σ', but experimental values of σ determined at low speeds for some gas-surface combinations are given in Table H,4b. Inspection of Tables H,4a and H,4b reveals that, for many combinations, α and σ are close to unity, which indicates that the molecules are reemitted nearly diffusely. It is thus to be expected that σ' will also be close to unity. In the calculations which follow, the two limiting cases of completely diffuse and completely specular interaction are presented.

Table H,4b. Values of reflection coefficient σ [25].

Gas and surface combination	σ
Air or CO_2 on machined brass or old shellac	1.00
Air on oil	0.895
CO_2 on oil	0.92
H_2 on oil	0.93
Air on glass	0 89
He on oil	0.87
Air on fresh shellac	0.79

It should be emphasized that the quantities α, σ, and σ' are over-all phenomenological averages. Additional interaction parameters could be defined. There is indication that α and possibly σ may depend on the surface temperature, gas temperature, and pressure; absorbed gas film characteristics [20, Pt. 2, Sec. 4]; and possibly the velocity and direction of the mean flow with respect to the surface [24]. Hence the tabulated values should be used with caution.

H,5. Free Molecule Flow Heat Transfer. Convective heat transfer calculations for a surface element in a steady, uniform, free molecule flow have been made by Sänger and Bredt [26], Luntz [27], and Stalder and Jukoff [28]. An energy balance for a differential surface element dA is written in the form

$$dQ = dE_i - dE_r \qquad (5\text{-}1)$$

where dQ is the total amount of heat removed from dA per unit time. If $\alpha \neq 0$, one obtains on elimination of the reflected energy flux with the help of Eq. 4-1,

$$dQ = \alpha(dE_i - dE_w) \qquad (5\text{-}2)$$

For the incident molecules, the energy flux dE_i is broken up into two components, the first $dE_{i,tr}$, which is due to the translational motion of the molecules, while the second $dE_{i,rot}$ is due to the internal degrees of freedom such as rotation or vibration. For the determination of $dE_{i,tr}$ one utilizes the assumption that the distribution of velocities for the incident molecules relative to an observer moving with the steady gas velocity U is Maxwellian. Then the number density of molecules per unit volume of phase space with total absolute velocity components v_{m_1}, v_{m_2}, and v_{m_3}, relative to the surface element shown in Fig. H,5, is given by [2]

$$f = \frac{\rho_\infty}{m(2\pi\Re T_\infty)^{\frac{3}{2}}} \exp\left[-\frac{(v_{m_1} - U\sin\theta)^2 + (v_{m_2} + U\cos\theta)^2 + v_{m_3}^2}{2\Re T} \right]$$

(5-3)

where m is the molecular mass, $\Re$ the specific gas constant, T_∞ and ρ_∞ the free stream temperature and density of the gas, and θ the angle of

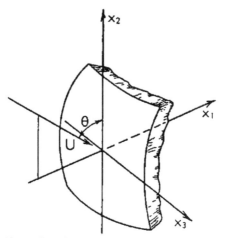

Fig. H,5.   Coordinate system for free molecule flow.

attack between the surface element dA and the flow direction. The flux of translational energy onto dA is then

$$dE_{i,tr} = \int_{-\infty}^{\infty} \int_{-\infty}^{\infty} \int_0^\infty \tfrac{1}{2} m v_m^2 v_{m_1} f dv_{m_1} dv_{m_2} dv_{m_3} dA$$

(5-4)

Integrating, one obtains

$$dE_{i,tr} = \rho\Re T \sqrt{\frac{\Re T}{2\pi}} \{(S^2 + 2)e^{-(S\sin\theta)^2}$$

$$+ \sqrt{\pi}\,(S^2 + \tfrac{5}{2})(S\sin\theta)[1 + \mathrm{erf}\,(S\sin\theta)]\}dA \quad (5\text{-}5)$$

where S denotes the "molecular speed ratio" which can also be expressed in terms of the Mach number M and the isentropic exponent γ by

$$S = \frac{U}{\sqrt{2\Re T}} = \sqrt{\frac{\gamma}{2}}\,M$$

(5-6)

The number of molecules N_i that are incident on dA per unit time is given by

$$N_i = \int_{-\infty}^{\infty} \int_{-\infty}^{\infty} \int_{0}^{\infty} v_{m_i} f dv_{m_i} dv_{m_2} dv_{m_3} dA \tag{5-7}$$

which on integration yields

$$N_i = \frac{\rho}{m} \sqrt{\frac{\Re T}{2\pi}} \{e^{-(S \sin \theta)^2} + \sqrt{\pi} \, (S \sin \theta)(1 + \text{erf} \, [S \sin \theta])\} dA \tag{5-8}$$

Each of these molecules carries by the principle of equipartition on the average $j(m\Re T/2)$ units of internal energy, where j denotes the number of internal degrees of freedom which partake in energy exchange with the surface; j is related to the appropriate γ by

$$j = \frac{5 - 3\gamma}{\gamma - 1} \tag{5-9}$$

The flux of internal energy incident on dA per unit time is hence

$$dE_{1, \text{int}} = \frac{5 - 3\gamma}{\gamma - 1} \frac{m\Re T}{2} N_i \tag{5-10}$$

The total flux of incident energy is then obtained by combining Eq. 5-5, 5-8, and 5-10.

By a similar line of reasoning one computes the energy flux dE_w in Eq. 5-2. This flux is due to particles which issue from the surface with no macroscopic velocity and in Maxwellian equilibrium at the temperature T_w. A comparison of Eq. 5-5 and 5-8 reveals that, for a gas at rest (i.e., $S = 0$) with a temperature T_w, the average translational energy carried by molecules across a unit area per unit time is $2m\Re T_w$. The reflected molecules may be thought of as having issued from such a hypothetical gas. Together with the appropriate internal energy contribution one then obtains

$$dE_w = (4 + j) \frac{m\Re T_w}{2} N_w = \frac{\gamma + 1}{2(\gamma - 1)} m\Re T_w N_w \tag{5-11}$$

where N_w is the number of particles reemitted from dA per unit time. Since for a steady state N_w must equal N_i, one obtains finally from Eq. 5-2, 5-5, 5-8, 5-10, and 5-11:

$$dQ = \alpha \rho \Re T \sqrt{\frac{\Re T}{2\pi}} \left(\left[S^2 + \frac{\gamma}{\gamma - 1} - \frac{\gamma + 1}{2(\gamma - 1)} \frac{T_w}{T} \right] \{e^{-(S \sin \theta)^2} \right.$$
$$\left. + \sqrt{\pi} \, (S \sin \theta)[1 + \text{erf} \, (S \sin \theta)]\} - \tfrac{1}{2}e^{-(S \sin \theta)^2} \right) dA \tag{5-12}$$

H,6. Heat Transfer Characteristics of Typical Bodies in Free Molecule Flow. Since dA is in general a function of θ, the basic relation (Eq. 5-12) may be integrated for any specific body shape in order to determine the total heat transfer Q. It is assumed in the following that the surface is everywhere convex towards the gas stream, which excludes the possibility of molecular inter-reflection.

In general the temperature T_w varies over the surface. However, if the thermal conductivity of the body is sufficiently large and the magnitude of heat convected to or from the surface is sufficiently small, T_w is very nearly constant. In the following it is assumed that this condition is always met.

The following two problems are of greatest interest, namely the determination of the recovery or equilibrium surface temperature T_r corresponding to the case of $Q = 0$ and the evaluation of the heat transfer Q to or from the body. In order to compare the results with continuum theory, the equilibrium and heat transfer characteristics are most conveniently expressed in terms of a modified thermal recovery factor r' and a modified Stanton number St' introduced by Oppenheim [29], both of which depend only on the speed ratio

$$r' = \frac{T_r - T_\infty}{T_\infty^0 - T_\infty} \frac{\gamma + 1}{\gamma}$$

$$St' = \frac{Q}{A\rho U c_p (T_r - T_w)} \frac{\gamma}{\alpha(\gamma + 1)}$$

(6-1)

T_∞^0 is the stagnation temperature of the flow which is given by

$$T_\infty^0 = T_\infty \left(1 + \frac{\gamma - 1}{\gamma} S^2\right)$$

A the total heat transfer surface area, and c_p the specific heat of the gas at constant pressure.

By direct integration of Eq. 5-12 one obtains the following results: For the flat plate, at angle of attack θ, with front and rear surface in perfect thermal contact and A the total area of both sides of the plate, one has

$$r' = \frac{1}{S^2} \left[2S^2 + 1 - \frac{1}{1 + \sqrt{\pi} (S \sin \theta) \text{ erf } (S \sin \theta) e^{(S \sin \theta)^2}}\right]$$

$$St' = \frac{1}{4 \sqrt{\pi} S} [e^{-(S \sin \theta)^2} + \sqrt{\pi} (S \sin \theta) \text{ erf } (S \sin \theta)]$$

(6-2)

For a flat plate, at angle of attack θ, with front and rear surfaces insulated from one another and A, the area of one side of the plate, one has for the front side

$$r' = \frac{1}{S^2}\left\{2S^2 + 1 - \frac{1}{1 + \sqrt{\pi}\,(S\sin\theta)[1 + \mathrm{erf}\,(S\sin\theta)]e^{(S\sin\theta)^2}}\right\} \quad (6\text{-}3)$$

$$St' = \frac{1}{4\sqrt{\pi}\,S}\left\{e^{-(S\sin\theta)^2} + \sqrt{\pi}\,(S\sin\theta)[1 + \mathrm{erf}\,(S\sin\theta)]\right\}$$

For the rear side one need only replace θ by $(-\theta)$ in these equations.

For the right circular cylinder with axis normal to the direction of flow and A the surface area without end contributions, one obtains from the calculations of Stalder, Goodwin, and Creager [15,16],

$$r' = \frac{(2S^2 + 3)I_0\left(\dfrac{S^2}{2}\right) + (2S^2 + 1)I_1\left(\dfrac{S^2}{2}\right)}{(S^2 + 1)I_0\left(\dfrac{S^2}{2}\right) + S^2 I_1\left(\dfrac{S^2}{2}\right)}$$

$$St' = \frac{e^{-S^2/2}}{4\sqrt{\pi}}\left[\frac{S^2 + 1}{S}I_0\left(\frac{S^2}{2}\right) + S I_1\left(\frac{S^2}{2}\right)\right] \quad (6\text{-}4)$$

where I_0 and I_1 are the modified Bessel functions. For a sphere with A the surface area one obtains from the calculations of Sauer [30]

$$r' = \frac{(2S^2 + 1)\left[1 + \dfrac{1}{S}\,\mathrm{ierfc}\,(S)\right] + \dfrac{2S^2 - 1}{2S^2}\,\mathrm{erf}\,(S)}{S^2\left[1 + \dfrac{1}{S}\,\mathrm{ierfc}\,(S)\right] + \dfrac{1}{2S^2}\,\mathrm{erf}\,(S)}$$

$$St' = \frac{1}{8S^2}[S^2 + S\,\mathrm{ierfc}\,(S) + \tfrac{1}{2}\,\mathrm{erf}\,(S)] \quad (6\text{-}5)$$

where ierfc (S) is the integrated complementary error function [31]. These results are exhibited in Fig. H,6a and H,6b.

A detailed investigation of the surface temperature of a flat plate at different angles of attack in the presence of radiation effects has been carried out by Stalder and Jukoff [28]. The calculations cover flight speeds up to 6.8 mi/sec and an altitude range of 75 to 190 miles.

It may be observed from Fig. H,6a that for all three bodies, and for the frontal side of the insulated flat plate, the recovery factors are always greater than unity. Hence, in contrast to a continuum flow, the equilibrium temperature in free molecule flow is greater than the local stagnation temperature of the gas. The anomaly can be explained by a consideration of the magnitudes of the incident and reemitted molecular energy fluxes. This striking effect has been verified experimentally by Stalder, Goodwin, and Creager for a cylinder in both a monatomic and diatomic gas [15,16].

It should be noted that the recovery factor, and hence also the equilibrium temperature T_r, are independent of the thermal accommodation

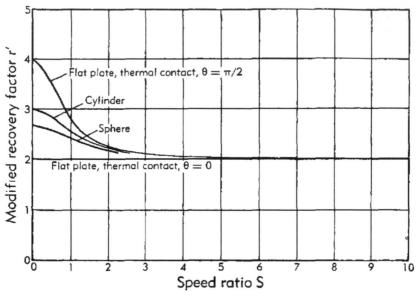

Fig. H,6a.   Modified recovery factor in free molecule flow.

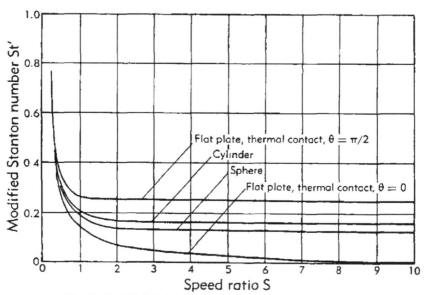

Fig. H,6b.   Modified Stanton number in free molecule flow.

coefficient. This is true if convection is the only mechanism of heat transfer present. The equilibrium surface temperature of a body obtained in the presence of convection, radiation, and internal conduction usually involves a balance between the three modes of energy transfer and is therefore dependent on α. On the other hand, if $\alpha = 0$ the energy exchange by convection ceases.

H,7. Aerodynamic Forces in Free Molecule Flow. Aerodynamic forces on typical body geometries have been calculated by many authors. The general method of attack has been nearly the same in all cases. A summary of the references is given in Table H,7. The force dF on the

Table H,7. Summary of references treating aerodynamic characteristics of bodies in free molecule flow.

Plate	Cylinder	Sphere	Cone	Ogive	Ellipsoid	Composite bodies
[4]	[27]	[27]	[27]	[33]	[32]	[33]
[17]	[32]	[32]	[32]	[34]	[37]	[34]
[27]	[34]	[33]	[33]	[36]		[37]
[32]	[35]	[34]	[34]	[37]		
[33]	[36]	[35]	[35]			
[34]	[37]		[37]			
[35]						

surface element dA consists of a component dF_i due to the incident molecules and a component dF_r due to reemitted molecules. It is convenient to resolve the stress dF_i/dA into a component p_i normal to the surface (pressure) and a component τ_i tangential to the surface (shear). With the help of the coordinate system shown in Fig. H,5 one obtains

$$p_i = \int_{-\infty}^{\infty} \int_{-\infty}^{\infty} \int_{0}^{\infty} mv_{m_1}^2 f dv_{m_1} dv_{m_2} dv_{m_3}$$

$$= \frac{\rho_\infty U^2}{2\sqrt{\pi}\, S^2} \{(S \sin \theta) e^{-(S \sin \theta)^2} \tag{7-1}$$

$$+ \sqrt{\pi} \left[\tfrac{1}{2} + (S \sin \theta)^2\right][1 + \text{erf}\,(S \sin \theta)]\}$$

$$\tau_i = \int_{-\infty}^{\infty} \int_{-\infty}^{\infty} \int_{0}^{\infty} mv_{m_1} v_{m_2} f dv_{m_1} dv_{m_2} dv_{m_3}$$

$$= -\frac{\rho_\infty U^2 \cos \theta}{2\sqrt{\pi}\, S} \{e^{-(S \sin \theta)^2} + \sqrt{\pi}\,(S \sin \theta)[1 + \text{erf}\,(S \sin \theta)]\} \tag{7-2}$$

where f is the molecular distribution function given by Eq. 5-3. From the definition of the reflection coefficients σ and σ' in Eq. 4-2 and 4-3, one has

$$p_r = (1 - \sigma')p_i + \sigma'p_w$$

$$\tau_r = (1 - \sigma)\tau_i \tag{7-3}$$

The net pressure p and shear τ are then given by

$$p = p_i + p_r = (2 - \sigma')p_i + \sigma'p_w$$

$$\tau = \tau_i - \tau_r = \sigma\tau_i \tag{7-4}$$

The pressure p_w exerted by molecules leaving the surface with no macroscopic velocity component and in Maxwellian equilibrium at temperature T_w is obtained with the help of Eq. 5-8 and 7-1. A comparison reveals that for a gas at rest (i.e. $S = 0$) with a temperature T_w the average normal momentum component carried by molecules across a unit area per unit time is $\frac{1}{2}m \sqrt{2\pi \Re T_w}$. The reflected particles may be thought of as having issued from such a hypothetical gas. Since for a steady state the number of incident molecules must equal the reflected number where N_i is given by Eq. 5-8, one has

$$p_w = \tfrac{1}{2}m \sqrt{2\pi \Re T_w}\, N_i \tag{7-5}$$

Combining the results of Eq. 7-1, 7-2, 7-3, 7-4, and 7-5 one obtains

$$p = \frac{\rho_\infty U^2}{2S^2}\left[\left(\frac{2-\sigma'}{\sqrt{\pi}} S \sin\theta + \frac{\sigma'}{2}\sqrt{\frac{T_w}{T}}\right) e^{-(S\sin\theta)^2}\right.$$

$$+ \left\{(2-\sigma')[(S \sin\theta)^2 + \tfrac{1}{2}]\right.$$

$$\left.\left. + \frac{\sigma'}{2}\sqrt{\frac{\pi T_w}{T}}\, (S \sin\theta)\right\} [1 + \mathrm{erf}\,(S \sin\theta)]\right] \tag{7-6}$$

$$\tau = -\frac{\sigma \rho_\infty U^2 \cos\theta}{2\sqrt{\pi}\,S} \{e^{-(S\sin\theta)^2} + \sqrt{\pi}\,(S \sin\theta)[1 + \mathrm{erf}\,(S \sin\theta)]\} \tag{7-7}$$

The total force acting on a body in a particular direction may be obtained by integrating the appropriate components of p and τ over the surface. Inspection of Eq. 7-6 and 7-7 reveals that the net forces in general depend on the surface temperature T_w (and hence α) as well as on both σ and σ'. In the following specific applications it is, for the sake of simplicity, assumed that the body is at a uniform temperature T_w equal to the free stream temperature and that either the reflection is completely diffuse, i.e. $\sigma = \sigma' = 1$ or that it is completely specular, i.e. $\sigma = \sigma' = 0$. In a more realistic calculation the surface temperature distribution must be determined from Eq. 5-12 taking the internal conduction and external radiation characteristics of the body into account. The force components should then be integrated over the surface with appropriate local values for α, σ, σ', and T_w. A number of investigations listed in Table H,7 have been concerned with the determination of the aerodynamic forces under the assumption, originally due to Maxwell, that a fraction σ of the incoming molecules is reflected diffusely and the remainder $(1 - \sigma)$ specularly. In this case Eq. 7-3 is replaced by

$$p = (2 - \sigma)p_i + \sigma p_r$$
$$\tau = \sigma \tau_i \tag{7-8}$$

where p_r is the pressure exerted by molecules leaving the surface in Maxwellian equilibrium at a temperature T_r which is not necessarily equal to the surface temperature T_w. T_r can be found with the help of the accommodation coefficient from Eq. 4-1.

H,8. Aerodynamic Force Characteristics of Typical Bodies in Free Molecule Flow. The basic relations given in Eq. 7-6 and 7-7 may be applied to a particular geometry in order to obtain by integration the lift and drag components acting on the body. The results are most conveniently expressed in terms of a drag coefficient C_D and a lift coefficient C_L. These are given by

$$C_D = \frac{D}{\frac{1}{2}\rho U^2 A}$$

$$C_L = \frac{L}{\frac{1}{2}\rho U^2 A} \tag{8-1}$$

where L and D are the total lift and drag forces respectively, and A is the reference area of the body which must be specifically defined in each case. By direct integration, the details of which may be found in the references listed in Table H,7, one obtains with the assumption of uniform surface temperature the following:

For a flat plate at an angle of attack θ and A the area of one side of a plate, one has for diffuse reflection

$$C_{D,\text{diff}} = \frac{2}{\sqrt{\pi} S}\left[e^{-(S \sin \theta)^2} + \sqrt{\pi}\, S \sin \theta \left(1 + \frac{1}{2S^2}\right) \text{erf}\, (S \sin \theta) \right.$$
$$\left. + \frac{\pi S}{S_w} \sin^2 \theta \right] \tag{8-2}$$

$$C_{L,\text{diff}} = \frac{\cos \theta}{S^2}\left[\text{erf}\,(S \sin \theta) + \sqrt{\pi}\, \frac{S^2 \sin \theta}{S_w} \right]$$

and for specular reflection,

$$C_{D,\text{spec}} = \frac{4 \sin \theta}{\sqrt{\pi} S^2}\, \{(S \sin \theta) e^{-(S \sin \theta)^2}$$
$$+ \sqrt{\pi}\, [\tfrac{1}{2} + (S \sin \theta)^2]\, \text{erf}\,(S \sin \theta)\} \tag{8-3}$$

$$C_{L,\text{spec}} = \cot \theta\, C_{D,\text{spec}}$$

For a right circular cylinder with its axis normal to the flow and A the projected area (no end contributions), one has for diffuse reflection:

$$C_{D,\text{diff}} = \frac{\sqrt{\pi}\, e^{-S^2/2}}{S}\left[\left(S^2 + \frac{3}{2}\right) I_0\left(\frac{S^2}{2}\right) + \left(S^2 + \frac{1}{2}\right) I_1\left(\frac{S^2}{2}\right) \right] + \frac{\pi^{\frac{3}{2}}}{4S_w} \tag{8-4}$$

and for specular reflection,

$$C_{D,\text{spec}} = \frac{4}{3}\left(C_{D,\text{diff}} - \frac{\pi^{\frac{3}{2}}}{4S_w} \right) \tag{8-5}$$

For a sphere with A the projected area, one has for diffuse reflection

$$C_{D.diff} = \frac{e^{-S^{1/2}}}{\sqrt{\pi}\,S^3}(1 + 2S^2) + \frac{4S^4 + 4S^2 - 1}{2S^4}\,\text{erf}\,(S) + \frac{2\sqrt{\pi}}{3S_w} \quad (8\text{-}6)$$

and for specular reflection,

$$C_{D.spec} = C_{D.diff} - \frac{2\sqrt{\pi}}{3S_w} \quad (8\text{-}7)$$

The quantity S_w which appears in these results is defined by

$$S_w = \frac{U}{\sqrt{2\Re T_w}} \quad (8\text{-}8)$$

and thus requires knowledge of the as-yet-unspecified surface temperature. The above results are shown in Fig. H,8a, H,8b, H,8c, and H,8d for the special case when the surface and free stream temperatures are

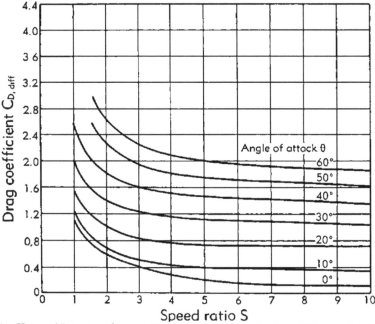

Fig. H,8a. Flat plate drag coefficient in free molecule flow. Diffuse reflection.

assumed equal, i.e. $S_w = S$. In this instance the equations do not contain γ and therefore apply to both monatomic and diatomic gases. A more realistic assumption can be made in the special case in which radiation and internal conduction effects are absent. Then the surface temperature equals the equilibrium temperature which can be obtained from the appropriate recovery factor in Art. 6. In this case the aerodynamic coefficients depend on γ and are somewhat larger than those for $S_w = S$, since the recovery factor is larger than unity.

The drag on a cylinder in free molecule flow has been determined experimentally by Stalder, Goodwin, and Creager [15] in satisfactory

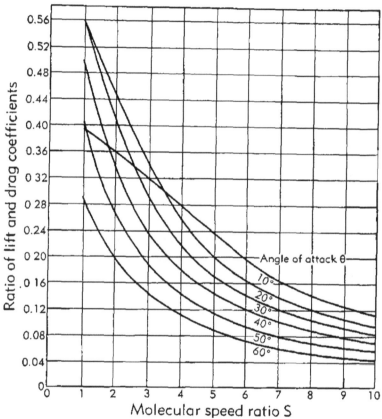

Fig. H,8b.   Flat plate lift-drag ratio. Diffuse reflection.

agreement with the theoretical results. The aerodynamic characteristics of flat plate airfoils has been the subject of a number of investigations (see Table H,7). Polar lift and drag diagrams indicate in general a poor flight efficiency under diffuse molecular reflection in contrast to the very high efficiency for specular reflecting surfaces.

H,9.   Nonuniform, Unsteady, and Surface-Interacting Free Molecule Flows.

The foregoing discussion was confined to uniform external flows which can be described by the Maxwellian equilibrium distribution given by Eq. 5-3. If the flow is no longer uniform due to the presence of

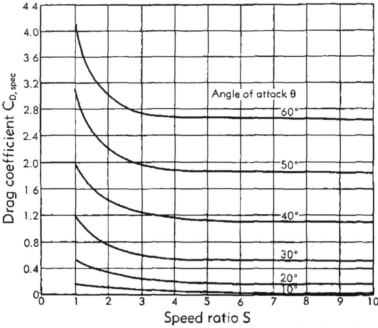

Fig. H,8c. Flat plate drag coefficient in free molecular flow. Specular reflection.

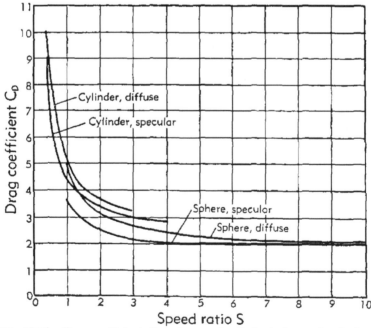

Fig. H,8d. Drag coefficients for sphere and cylinder in free molecule flow.

shear stresses and heat flux components one may use as a first approximation a distribution function proposed by Maxwell [1, p. 681], and
given in Eq. 12-9. The methods of the previous articles have been applied
to the calculation of the heat transfer and the aerodynamic forces on a
cylinder [38,39] for this case. It is found that the effect on the pertinent
parameters due to the nonuniformity is in general small compared to a
uniform flow, except at low speeds or for flows with very large gradients.

Tsien [5] first suggested that a cylinder of sufficiently small diameter
could serve as a temperature probe in a free molecule flow. From a cor-
rected reading of the equilibrium temperature and the knowledge of the
local stagnation temperature one can then determine the Mach number of
the stream. Such an instrument has recently been used by Sherman [40]
in a study of the structure of a shock wave where the nonuniformity of
the flow affects the reading of the device.

A case of unsteady motion in a free molecule flow has been treated
by Burgers [41]. The relevant feature of the problem is that gas is in contact
with a surface which has a large variable molecular speed ratio. The
state of the gas at any point, at any instant, is dependent in part on those
molecules which left the surface at an earlier time. Their number and
energy is determined by the surface velocity at the moment of their
reflection. Consequently, even with completely diffuse reemission an
appreciable deviation from the Maxwellian distribution appears in the
reflected gas.

For very large molecular speed ratios there arises the possibility that
the flow may begin to cause ablation from the surface of the body. This
is in part due to the fact that the heat transfer to a rapidly moving object
becomes so great that surface melting or vaporization may occur, and in
part due to the mechanical action of the aerodynamic forces. Under these
extreme conditions one may consider the molecules essentially at rest
and the transfer of momentum and energy due to those particles which
are intercepted by the body in its travel. If the reflected molecular contributions are considered negligible compared to the incoming molecular
flux, Eq. 5-12, 7-1, and 7-2 reduce to

$$Q = \alpha \tfrac{1}{2}\rho U^3 A \sin \theta \tag{9-1}$$

and [13]

$$p_i = \tfrac{1}{2}\rho U^2 2 \sin^2 \theta$$
$$\tau_i = \tfrac{1}{2}\rho U^2 2 \sin \theta \cos \theta \tag{9-2}$$

where $A \sin \theta$ is the projected cross section of the body perpendicular to
the direction of flight. Thomas and Whipple [42] have summarized and
correlated the available experimental heat transfer data in terms of the
parameter $\Lambda = Q/\tfrac{1}{2}\rho U^3 S'$ where S' is the effective heat transfer area of

the body. From meteor observations up to speeds of 40 mi/sec, where considerable surface ablation occurs, they conclude that $\Lambda \lesssim 0.1$ and that it decreases with decreasing speed. If the velocity becomes sufficiently high to melt the surface material and if radiation effects are negligible one can estimate the mass loss per unit time per unit area $= \Lambda \frac{1}{2}\rho U/L$ where L is the heat of vaporization or fusion according to which process the mass loss is due. Eq. 9-2 is the well-known Newton resistance law for inelastic impact. Meteor observations [42] indicate that the drag coefficient based on the projected area perpendicular to the direction of flight is of the order of one half.

Sänger [43] has made some estimates of the effects on the energy transfer to the surface due to excitation, dissociation, and ionization of molecules which might be produced by extremely high impact speeds between particles and the body. In general these effects decrease the energy transfer to the surface due to a partial transfer of the energy into higher molecular energy states. For air, Sänger estimates that only elastic reflection begins to be affected at speeds larger than 1.24 mi/sec.

H,10. Transition Regime Calculations. The foregoing discussion has been confined to the case of free molecule flow in which the effects of molecular encounters have been entirely neglected. With increasing density the effect of these collisions begins to be of importance and the low density part of the transition regime is entered. The general formulation of the problem in this regime is due to Jaffé [44] and is based on a perturbation expansion of the Maxwell-Boltzmann equation in inverse powers of the molecular mean free path. The distribution function f, which satisfies this equation, is assumed to be of the form

$$f = f_0 \left[1 + \frac{L}{l} \varphi_1 + \left(\frac{L}{l} \right)^2 \varphi_2 + \cdots \right] \tag{10-1}$$

where f_0 is the Maxwellian equilibrium distribution and L a typical macroscopic dimension.

It is no longer sufficient to specify certain over-all boundary conditions on the energy and momentum fluxes as presented in Eq. 4-1, 4-2, and 4-3. One must in addition specify conditions on f itself. A number of special calculations have been carried out in this regime by Heineman [32], Keller [45], Wang-Chang [46], Wang-Chang and Uhlenbeck [47], Szymanski [48], Lunc [49], Lunc and Lubonski [50], and Kryzwoblocki [51]. The calculations are quite formidable and no results of direct aerodynamic interest are available. Experimental data obtained by Bowyer [52] provide confirmation of the theoretical predictions [46] for the torque transmitted between concentric rotating cylinders in this part of the transition regime.

CHAPTER 3. SLIP FLOW

H,11.   Introduction.   The slip flow regime is the flow regime of slight rarefaction. The gas density is just slightly less than that characteristic of a completely continuum flow. In the slip flow regime there are really three separately important, but interrelated, parameters, the Mach number M, the Reynolds number Re, and the appropriate Knudsen number, either M/Re or $M/\sqrt{Re}$. These parameters serve to indicate the importance of compressibility, viscosity, and rarefaction effects, respectively. For the Knudsen number to be in the range indicated in Eq. 2-3, i.e. of the order of 0.01 to 0.1, it is clear that either M must be large, or Re small, (or both). Hence rarefaction effects in the slip flow regime are associated with, and in fact often dominated by, very strong compressibility or viscosity effects. In general, it is to be expected that boundary layers will be laminar; they will often be quite thick and in fact the Reynolds number may be so low that boundary layer theory is not strictly applicable. It is also to be expected that interaction effects between these thick viscous layers and the supersonic inviscid flow field [53,54,55] will be of considerable importance. Because of the complexity of these interrelations in the slip flow regime there are few, if any, solutions yet available for specific flow situations which take proper account of viscosity, compressibility, and rarefaction effects. The general effect of the slip flow and temperature jump boundary conditions of Eq. 14-3 is to reduce skin friction and heat transfer, other factors being similar.

There is a considerable and rapidly growing body of experimental data of aerodynamic interest covering portions of the slip flow as well as the transition flow regimes. Some of this information can be discussed in terms of comparatively approximate theories while some of it is only empirical. This material is summarized in Art. 15 to 21. However, the basic and fundamental problem of the correct formulation of the proper differential equations and boundary conditions for describing slip flows in general is not yet resolved in a satisfactory way. Until recently, it had been generally considered that some modification of the Navier-Stokes equations, such as the Burnett equations [4] or Thirteen Moment equations [56], or those proposed by Koga [57] together with slip velocity and temperature jump boundary conditions (see Art. 14) could be regarded as applicable to the slip flow regime. This view has been held in spite of several well-recognized difficulties associated with the Burnett and Thirteen Moment equations, and also in spite of the lack of any experimental verification of their validity. Recent experimental and theoretical findings, although not yet completely conclusive, now make the validity of these equations appear less likely and in fact seem to indicate that the Navier-Stokes equations, together with slip velocity and temperature jump boundary conditions, are not only adequate but probably superior. In the next article, a very brief derivation of the Thirteen Moment and Burnett equations is given, after which some of the difficulties met are discussed.

H,12. The Thirteen Moment and Burnett Equations. Both of these sets of equations are intended as approximations to the basic Maxwell-Boltzmann equation (Eq. 12-4) of the kinetic theory of monatomic gases. Each represents the third step in a successive approximation scheme. The Thirteen Moment equations were implicitly indicated by Maxwell [1, p. 681] but the method seemed applicable only to "Maxwell molecules," i.e. molecules which repelled each other according to the inverse fifth power of their distance of separation. A method of solution applicable to any molecular model was suggested by Hilbert, and calculations were carried out by Enskog [58], Burnett [59], Chapman and Cowling [60], and others. More recently, Grad [56] has completed the calculations as formulated by Maxwell and obtained the Thirteen Moment equations for the Maxwell molecule. It has also been shown [61, p. 237] that the Thirteen Moment equations contain the Burnett equations as a special case for Maxwell molecules. The Burnett equations for other molecular models differ only slightly from those for Maxwell molecules. These results are presented in complete details in [56; 60; 62, Chap. 7; 63], but the essential elements are briefly recalled here.

The gas is supposed to consist of molecules of mass m, having only three degrees of freedom, so that the state of the gas may be given in terms of a distribution function,

$$f(x_i, v_{m_i}, t), \qquad i = 1, 2, 3 \tag{12-1}$$

which is the number density of molecules at point x_i with velocity v_{m_i} at time t, within the phase space volume element,

$$dx dv_m = dx_1 dx_2 dx_3 dv_{m_1} dv_{m_2} dv_{m_3} \tag{12-2}$$

The ordinary macrosopic variables of gas dynamics—the density ρ, the velocity u_i, the stress τ_{ij} the temperature T, and the heat flux q_i—are given in terms of f by

$$\rho = m \int_{-\infty}^{\infty} \int_{-\infty}^{\infty} \int_{-\infty}^{\infty} f dv_{m_1} dv_{m_2} dv_{m_3} \equiv m \int f dv_m \equiv mn$$

$$u_i = \frac{1}{n} \int v_{m_i} f dv_m$$

$$P_{ij} = m \int (v_{m_i} - u_i)(v_{m_j} - u_j) f dv_m \equiv m \int V_i V_j f dv_m \equiv p\delta_{ij} - \tau_{ij} \tag{12-3}$$

$$T = \frac{1}{n\Re} \int V^2 f d\mathbf{v}_m = \frac{p}{\Re\rho}$$

$$q_i = \tfrac{1}{2}m \int V_i V^2 f d\mathbf{v}_m$$

where $\Re$ is the specific gas constant.

The distribution function obeys a conservation equation, i.e. the Maxwell-Boltzmann equation,

$$\frac{\partial f}{\partial t} + \sum_{i=1}^{3} \left(v_{m_i} \frac{\partial f}{\partial x_i} + X_i \frac{\partial f}{\partial v_{m_i}} \right) = \Delta_c f \qquad (12\text{-}4)$$

where X_i is an external force field and $\Delta_c f$ is the time rate of change of f due to intermolecular collisions. The term $\Delta_c f$ is given by

$$\Delta_c f = \iiint d\mathbf{v}_{m_1} \iint d\Omega[gI(g,\theta)(f'f_1' - ff_1)] \qquad (12\text{-}5)$$

where $g = |\mathbf{v}_{m_1} - \mathbf{v}_m|$ is the relative velocity of the given molecule and a colliding molecule whose velocity is $\mathbf{v}_{m_1}$, the quantity $gI(g,\theta) d\Omega$ is the probability that the colliding molecule will be scattered into $d\Omega$ at θ, and f' and f_1' denote the distribution functions for molecules after collision, with velocities $\mathbf{v}_m'$ and $\mathbf{v}_{m_1}'$. Multiplying Eq. 12-4 by m, v_{m_j}, and V^2, respectively, and integrating over all velocities furnishes the usual macroscopic flow equations (assuming $X_i \equiv 0$):

$$\frac{\partial \rho}{\partial t} + \sum_{i=1}^{3} \frac{\partial(\rho u_i)}{\partial x_i} = 0$$

$$\frac{\partial u_i}{\partial t} + \sum_{j=1}^{3} \left(u_j \frac{\partial u_i}{\partial x_j} + \frac{1}{\rho} \frac{\partial P_{ij}}{\partial x_j} \right) = 0 \qquad (12\text{-}6)$$

$$\frac{\partial p}{\partial t} + \sum_{i=1}^{3} \left(\frac{\partial(pu_i)}{\partial x_i} + \frac{2}{3} \frac{\partial q_i}{\partial x_i} + \frac{2}{3} \sum_{j=1}^{3} P_{ij} \frac{\partial u_i}{\partial x_j} \right) = 0$$

The collision term disappears identically. This system of equations is incomplete since there are more unknowns ρ, p, u_i, τ_{ij} (13 in all) than equations (5 in all). The Hilbert-Enskog-Chapman-Burnett method obtains the necessary additional relations in terms of a "solution" of Eq. 12-4. It is assumed that the collision term is dominant and that f may be determined by successive iterations on the collision term. This is roughly equivalent to supposing that f is given in terms of a power series in the mean free path $\bar{l}$. Eq. 12-4 is reduced to a sequence of linear integral equations, which are in turn solved approximately by replacing them by systems of simultaneous linear equations. The first approximation to f is the equilibrium distribution

$$f^{(0)} = \frac{n}{(2\pi\Re T)^{\frac{3}{2}}}\, e^{-\frac{V^2}{2\Re T}} \tag{12-7}$$

The substitution of Eq. 12-7 gives the Euler equations of ideal-fluid theory. Similarly, the second approximation $f^{(1)}$ gives the Navier-Stokes equations, while the third approximation $f^{(2)}$ gives the Burnett equations. It is a consequence of the iteration method of solution that f, and hence τ_{ij} and q_i, are given, apparently exactly if the calculation were continued, in terms of space gradients of ρ, u_i, and T only. This is a form of the "Hilbert paradox" and indicates the special nature of the solutions of Eq. 12-4, to which the iteration method is restricted. Distribution functions which are not of this form are simple to formulate. For example, a distribution function for which there is a time-dependent viscous shear stress, $\tau(t)$, with *no* space variation of the velocity, is given by

$$f = f^{(0)}\left[1 + \frac{V_1 V_2}{2p\Re T}\tau(t)\right] \tag{12-8}$$

To obtain the Thirteen Moment equations, one starts with the substitution of

$$f = f^{(0)}\left[1 - \sum_{i=1}^{3}\sum_{j=1}^{3}\frac{\tau_{ij}}{2p\Re T}\,V_i V_j - \sum_{i=1}^{3}\frac{q_i}{p\Re T}\,V_i\left(1 - \frac{V^2}{5\Re T}\right)\right] \tag{12-9}$$

into Eq. 12-4. Multiplying successively by each of the fourteen quantities m, v_{m_i}, $\frac{1}{2}mV^2$, $V_i V_j$, $\frac{1}{2}V_i V^2$, and integrating over all molecular velocities, yields Eq. 12-6 and eight additional independent equations in the form ($\bar{A}_{ij}$ denotes $A_{ij} + A_{ji} - \frac{2}{3}\delta_{ij}\sum_{k=1}^{3}A_{kk}$):

$$\frac{\partial \tau_{ij}}{\partial t} + \sum_{k=1}^{3}\frac{\partial}{\partial x_k}(u_k\tau_{ij}) - \frac{2}{5}\overline{\frac{\partial q_i}{\partial x_j}} + \sum_{k=1}^{3}\overline{\tau_{ik}\frac{\partial u_j}{\partial x_k}} - p\overline{\frac{\partial u_i}{\partial x_j}} = -\frac{p}{\mu}\tau_{ij}$$

$$\frac{\partial q_i}{\partial t} + \sum_{k=1}^{3}\left[\frac{\partial(u_k q_i)}{\partial x_k} + \frac{7}{5}q_k\frac{\partial u_i}{\partial x_k} + \frac{2}{5}q_k\frac{\partial u_k}{\partial x_i} + \frac{2}{5}q_i\frac{\partial u_k}{\partial x_k}\right. \tag{12-10}$$

$$\left. - \Re T\frac{\partial \tau_{ik}}{\partial x_k} - \frac{7}{2}\tau_{ik}\Re\frac{\partial T}{\partial x_k} + \sum_{j=1}^{3}\frac{\tau_{ij}}{\rho}\frac{\partial P_{jk}}{\partial x_k}\right] + \frac{5}{2}p\Re\frac{\partial T}{\partial x_i} = -\frac{2}{3}\frac{p}{\mu}q_i$$

These equations are exact for "Maxwell" molecules, i.e. molecules which repel each other with a force proportional to the inverse fifth power of the distance of separation. For other models, certain very small additional terms are present which have been ignored above. The terms,

$$-\frac{p}{\mu}\tau_{ij} \quad \text{and} \quad -\frac{2}{3}\frac{p}{\mu}q_i \tag{12-11}$$

are obtained from the collision term. All other terms arise in the convection terms. The quantity μ is identified as the coefficient of viscosity and is given by

$$\mu = 0.243\left(\frac{2m}{A}\right)^{\frac{1}{2}} m\mathcal{R}T \qquad \text{Maxwell molecule } (F \sim r^{-5}) \tag{12-12}$$

and

$$\mu = \tfrac{5}{8}m\sqrt{\pi\mathcal{R}T}\,\frac{1}{2\pi d^2} \qquad \text{hard sphere, diameter } d$$

The system of 13 equations, i.e. Eq. 12-6 and 12-10, constitutes a sufficient system for the determination of the 13 unknowns listed previously. It is observed that τ_{ij} and q_i do not depend in any simple way on ρ, u_i, and T, or their gradients of any order. For a system which is uniform in space, Eq. 12-10 reduces, for example, to

$$\frac{\partial \tau_{ij}}{\partial t} + \frac{p}{\mu}\tau_{ij} = 0 \tag{12-13}$$

This has the solution

$$\tau_{ij} \sim e^{-pt/\mu} \tag{12-14}$$

The quantity μ/p is then a "relaxation" time (of the order of the time between successive collisions) for the disappearance of the stress.

The Euler, Navier-Stokes, and Burnett equations are all contained in the system of Eq. 12-6 and 12-10 if it is assumed that τ_{ij} and q_i can be given in terms of successive approximations by iteration on the collision term in Eq. 12-10. One has, for the nth approximation to the stress tensor and heat flux vector,

$$
\begin{aligned}
\tau_{ij}^{(n)} &= \mu\frac{\overline{\partial u_i}}{\partial x_j} - \frac{\mu}{p}\left[\frac{D\tau_{ij}^{(n-1)}}{Dt} + \tau_{ij}^{(n-1)}\sum_{k=1}^{3}\frac{\partial u_k}{\partial x_k}\right.\\
&\qquad\qquad \left. - \frac{2}{5}\overline{\frac{\partial q_i^{(n-1)}}{\partial x_j} + \sum_{k=1}^{3}\tau_{ik}^{(n-1)}\frac{\partial u_j}{\partial x_k}}\right]\\
q_i^{(n)} &= -k_{\mathrm{th}}\frac{\partial T}{\partial x_i} - \frac{3\mu}{2p}\left(\frac{Dq_i^{(n-1)}}{Dt} + \sum_{j=1}^{3}\left\{\tfrac{7}{5}q_i^{(n-1)}\frac{\partial u_j}{\partial x_j}\right.\right. \\
&\qquad - \mathcal{R}T\frac{\partial \tau_{ij}^{(n-1)}}{\partial x_j} + q_j^{(n-1)}\left[\frac{\partial u_i}{\partial x_j} + \frac{2}{5}\left(\frac{\partial u_i}{\partial x_j} + \frac{\partial u_j}{\partial x_i}\right)\right] \\
&\qquad \left.\left. - \tfrac{7}{5}\tau_{ij}^{(n-1)}\mathcal{R}\frac{\partial T}{\partial x_j} + \sum_{k=1}^{3}\frac{\tau_{ij}^{(n-1)}}{\rho}\frac{\partial P_{jk}^{(n-1)}}{\partial x_k}\right\}\right)
\end{aligned} \tag{12-15}
$$

⟨ 29 ⟩

where $k_{th} = 15 \Re \mu/4$ is identified as the coefficient of thermal conductivity, and D/Dt is the material derivative.

Beginning with $p_{ij}^{(0)} = q_i^{(0)} = 0$, i.e. the Euler equations, one obtains as the next approximation

$$\tau_{ij}^{(1)} = \mu \overline{\frac{\partial u_i}{\partial x_j}}$$

$$q_i^{(1)} = -k_{th} \frac{\partial T}{\partial x_i}$$

(12-16)

These yield the Navier-Stokes equations (for a monatomic gas) when substituted into Eq. 12-6. As the next step, one obtains

$$\tau_{ij}^{(2)} = \mu \overline{\frac{\partial u_i}{\partial x_j}} - \frac{\mu}{p}\left[\frac{D}{Dt}\left(\mu \overline{\frac{\partial u_i}{\partial x_j}} \right) + \mu \overline{\frac{\partial u_i}{\partial x_j}} \sum_{k=1}^{3} \frac{\partial u_k}{\partial x_k} \right.$$

$$\left. + \frac{2}{5}\overline{\frac{\partial}{\partial x_j}\left(k_{th} \frac{\partial T}{\partial x_i} \right)} + \mu \sum_{k=1}^{3} \overline{\frac{\partial u_i}{\partial x_k}\frac{\partial u_j}{\partial x_k}} \right]$$

$$q_i^{(2)} = -k_{th}\frac{\partial T}{\partial x_i} + \frac{3\mu}{2p}\left\{ \frac{D}{Dt}\left(k_{th}\frac{\partial T}{\partial x_i} \right) + \sum_{j=1}^{3}\left[\frac{7}{5}k_{th}\frac{\partial T}{\partial x_i}\frac{\partial u_j}{\partial x_j} \right.\right.$$

(12-17)

$$+ \Re T \frac{\partial}{\partial x_j}\left(\mu \overline{\frac{\partial u_i}{\partial x_j}} \right) + k_{th}\frac{\partial T}{\partial x_j}\left[\frac{\partial u_i}{\partial x_j} + \frac{2}{5}\left(\frac{\partial u_i}{\partial x_j} + \frac{\partial u_j}{\partial x_i} \right) \right]$$

$$\left.\left. + \frac{7}{2}\mu \overline{\frac{\partial u_i}{\partial x_j}} \Re \frac{\partial T}{\partial x_j} - \frac{\mu}{p}\sum_{k=1}^{3}\overline{\frac{\partial u_i}{\partial x_j}\frac{\partial p}{\partial x_k}} \delta_{jk} \right]\right\}$$

These yield the Burnett equations (for Maxwell molecules) when substituted into Eq. 12-6. For non-Maxwell molecules, some of the numerical coefficients in Eq. 12-17 are slightly different, but that is the only difference. It can be shown [4] that the ratios $(\tau_{ij}^{(2)} - \tau_{ij}^{(1)})/\tau_{ij}^{(1)}$ and $(q_i^{(2)} - q_i^{(1)})/q_i^{(1)}$ are in general proportional to M^2/Re so that the higher order approximations would become of importance in the slip flow regime, provided, of course, that they are correct.

It is of historical interest to note that those parts of the expression for $\tau_{ij}^{(2)}$ given in Eq. 12-17, which depend on temperature gradients, were obtained by Maxwell [1, p. 681] in order to explain the radiometer force in the slip flow regime. This effect is now thought to be due rather to the "thermal creep" boundary effect [61, p. 333] and explainable in terms of the Navier-Stokes stresses only.

H,13. Difficulties Associated with the Thirteen Moment and Burnett Equations. Some of the basic difficulties associated with these equations are as follows:

1. They are confined to monatomic gases. Since air is composed, for the most part, of diatomic nitrogen and oxygen, this is a serious restriction for aerodynamic applications. At normal temperatures, the internal energy in air is distributed equally over the translational and rotational degrees of freedom of the diatomic molecules, while the vibrational degrees of freedom are almost completely unexcited. Following any abrupt change in the state of the gas, e.g. upon passage through a shock wave, the internal energy is redistributed over the translational and internal degrees of freedom of the molecules, and a new equilibrium is achieved only after a sufficient number of molecular collisions, as indicated in Eq. 12-14. It seems reasonably well established experimentally [40,64,65,66] that the rotational degrees of freedom for air adjust in a relaxation time which corresponds to only a very few additional collisions. Under these circumstances, Wang-Chang and Uhlenbeck [67] have shown that the only macroscopically apparent effect—at least to the Navier-Stokes level of approximation—is the phenomenon of "bulk viscosity," i.e. the stress is given by

$$\tau_{ij} = \mu \overline{\frac{\partial u_i}{\partial x_j}} + \lambda \delta_{ij} \sum_{k=1}^{3} \frac{\partial u_k}{\partial x_k} \qquad (13\text{-}1)$$

The experiments seem to indicate that, for air, the magnitude of the second coefficient of viscosity λ is about $\frac{2}{3}$ the ordinary shear modulus μ. At higher temperatures, in air, the vibrational degrees of freedom begin to be excited and the readjustments of the internal energy associated with these modes of motion have very much longer relaxation times. Then, no simple formulas, as given by Eq. 13-1, are valid and phenomena become very complicated.

Almost nothing is known of such relaxation phenomena to the next order of approximation. Truesdell [68,69] has obtained general formulas for the stress tensor and heat flux vector, subject only to the necessary restrictions of dimensional and geometrical invariance, but which follow from the basic—and crucial—assumption that these quantities are given in terms of derivatives of the velocity, pressure, and temperature. Truesdell's formulas contain many more terms, at the Burnett level, than are given in Eq. 12-17. Possibly the missing terms can be associated with such relaxation effects. The kinetic theory approach to the problem has been formulated by Wang-Chang and Uhlenbeck [67], who have shown the relation of the second viscosity coefficient to the relaxation effect and have indicated the necessary calculations for determining higher order terms. A "Seventeen Moment" set of equations, including a second "temperature" associated with the internal degrees of freedom and the three components of the corresponding heat flux vector, has been formulated by Mostov, et al. [70] for a molecular model consisting of a rough sphere. This model has also been considered by Kohler [71]. These investigations also exhibit bulk viscosity phenomena.

In summarizing, it is clear that Eq. 12-17 are not complete for air, even to the Navier-Stokes level of approximation.

2. Present experimental evidence seems contrary to either the Burnett or Thirteen Moment equations. The crucial experiment to test the validity of either of these sets of equations must be carried out for a flow system having large gradients far away from any surface so as to avoid any ambiguities connected with boundary conditions or wall interference. Two such experiments are the determination of the internal structure of weak normal shock waves, and the determination of the dispersion and absorption of ultrasonic sound waves. The former has been studied by Greene and Andersen [64,72] and Sherman and Talbot [40,73], while the latter has been investigated most extensively by Greenspan [65]. Solutions are obtainable for the Navier-Stokes, Burnett, and Thirteen Moment equations for both cases.

The initial shock structure studies [40] were carried out in air and in helium at shock strengths above $M = 1.8$. Here they showed complete agreement with the Navier-Stokes prediction (including a bulk viscosity term for the case of air). At this shock strength the Burnett profile is quite different from the Navier-Stokes profile and the Thirteen Moment equations cannot be solved. The more recent shock structure studies [73] were carried out in argon at lower Mach numbers. For these conditions, all three sets of equations may be solved; the experimental results were in best agreement with the Navier-Stokes relations, particularly in the initial part of the shock wave which was undisturbed by downstream expansions. The sound dispersion and absorption experiments using an acoustic interferometer have not been completely free from wall interference and the results are not conclusive; they seem in better agreement with the Navier-Stokes equations, however, than with the other sets.

3. Recent theoretical results also seem contrary to the Burnett or Thirteen Moment equations. Several "exact" solutions of the Boltzmann equation itself have been obtained, which do not seem to be in accord with the basic assumptions underlying the Burnett or Thirteen Moment formulations. The linearized Boltzmann equation has been solved by Wang-Chang and Uhlenbeck for the case of heat transport between parallel plates [47] and for the case of flow near a surface [74]. The molecular distribution function f has the property for both these cases that $f \rightarrow f^0$ as $l \rightarrow 0$, but f is not expandable in a power series in l. Functions of the type $e^{-z/l}$ occur. The case of simple shear in an infinite medium has

been solved by Ikenberry and Truesdell [75,76] in terms of the complete Boltzmann equation. Here again the distribution function cannot be adequately approximated by either the Burnett or Thirteen Moment representations. For all of these theoretical cases it is noteworthy that the Navier-Stokes results are in general a better approximation than either the Burnett or Thirteen Moment results.

4. Even if the Burnett equations were valid, there is an additional and unresolved problem connected with the formulation of proper boundary conditions. The order of the Burnett equations is higher than the order of the Navier-Stokes equations, so that additional boundary conditions seem necessary. At the Euler level of approximation the equations are of first order, and only one boundary condition, usually that the normal velocity vanish at a surface, can be prescribed. At the Navier-Stokes level of approximation, the equations are of second order and two boundary conditions, usually that the normal and the tangential velocity vanish at the surface, must be prescribed. Solutions are thus singular in the viscosity (or equivalently, the mean free path), and the boundary layer solution (for $\nu \to 0$) cannot be obtained by perturbation schemes starting with the inviscid solution. At the Burnett level of approximation, the equations are of third order and it is to be expected that an additional boundary condition must be prescribed, although there is no clear guide as to which one. Schamberg [77] has suggested that solutions which are accurate to the Burnett level may be obtained by perturbation series techniques starting with the Navier-Stokes solution. If this is the case no additional conditions are necessary, or in fact, can be used and the question would be only one of improving the form of the boundary conditions (to include slip and temperature jump effects). Although the Euler-to-Navier-Stokes equivalent of this suggestion definitely does not work, and in fact, completely loses the whole boundary layer phenomenon, only a solution obtained directly from the Maxwell-Boltzmann equation could provide confirmation, or the reverse, of this hypothesis. The solution obtained by Wang-Chang and Uhlenbeck [47] may be reduced to macroscopic terms to give the heat flux in terms of the temperature drop and the mean free path $\bar{l}$. For small $\bar{l}$, the heat flux agrees with that predicted by the Navier-Stokes equations and the simple temperature jump boundary condition of Eq. 14-3, below. The correction term, however, is nonanalytic in $\bar{l}$ and so a perturbation calculation will not work. Although these results thus strongly indicate that Schamberg's suggestion will not work, they do not definitely prove it because, among other things, the Burnett terms disappear identically for this case. For the Thirteen Moment equations, such boundary condition uncertainties do not occur. Grad [56] has shown that the characteristic curves for these equations are such as to require the same number of boundary conditions as for the Navier-Stokes equations. Hence approximating the Thir-

teen Moment equation by the Navier-Stokes equations can be done unambiguously.

In conclusion, it seems best at present to rely on the Navier-Stokes equations, including a second viscosity coefficient, together with the slip velocity and temperature jump boundary conditions given in Eq. 14-3, to provide the basic system for the slip flow regime. At lower densities, in the beginning of the transition regime, the theoretical results suggest that such solutions may remain fairly good, but that any corrections must be of a very complicated nature, possibly involving complete solutions of the Maxwell-Boltzmann equation. The Navier-Stokes equations of course do not remain valid at very low densities, well into the transition regime; however, the Burnett and Thirteen Moment equations do not seem able even to predict when the Navier-Stokes relations break down.

H,14. Slip Velocity and Temperature Jump Boundary Conditions.   It has been recognized since the time of Smoluchowski [*18,78*] and Knudsen [*19*] that there is a slip velocity and a temperature jump at the interface between a solid and a gas at low pressures. An approximate analysis in a monatomic gas adjacent to an isothermal surface serves to relate the slip velocity and the gradient of the tangential velocity. Near the wall the gas consists of molecules, one half of which have just come off the wall, the other half of which have come, on the average, from a layer of gas a mean free path away. If y is the normal coordinate, one has

$$u(0) = \tfrac{1}{2} \left(\left[u(0) + l\left(\frac{\partial u}{\partial y}\right)_0 \right] \right.$$
$$\left. + \left\{ (1 - \sigma)\left[u(0) + l\left(\frac{\partial u}{\partial y}\right)_0 \right] + \sigma \cdot 0 \right\} \right) \quad (14\text{-}1)$$

where σ is the fraction of diffusely reflected molecules (whose average tangential velocity is thus 0) and $1 - \sigma$ is the fraction of specularly reflected molecules (whose average tangential velocity is the same as that of the molecules incident from the layer a distance l above the wall). Hence

$$u(0) = \frac{2 - \sigma}{\sigma} l\left(\frac{\partial u}{\partial y}\right)_0 \quad (14\text{-}2)$$

It is observed that, even for completely diffuse reflection, i.e. $\sigma = 1$, there is a definite slip velocity. At normal densities l is so small that the effect is completely negligible. Very much more elegant calculations, involving approximate solutions of the Maxwell-Boltzmann equation in the vicinity of a wall have been carried out by many investigators, e.g.

[22,23,56,74,77,79,80,81,82]. The results given by Kennard [2] are (within a numerical factor close to unity):

$$u(0) = \frac{2 - \sigma}{\sigma} \bar{l} \left(\frac{\partial u}{\partial y}\right)_0 + \frac{3}{4} \frac{\mu}{\rho T} \left(\frac{\partial T}{\partial x}\right)_0,$$

$$T(0) - T_w = \frac{2 - \alpha}{\alpha} \frac{2\gamma}{\gamma + 1} \frac{\bar{l}}{Pr} \left(\frac{\partial T}{\partial y}\right)_0,$$

(14-3)

where α is accommodation coefficient defined in Eq. 4-1.

There is some disagreement with the temperature jump condition for polyatomic gases, and Welander [81] and Wang-Chang and Uhlenbeck [74] also disagree with the factor $(2 - \alpha)/\alpha$. For macroscopic considerations these differences are not serious since they only correspond to different macroscopically determined values for the coefficients σ and α. The values tabulated in Table H,4a and H,4b for these quantities are in agreement with the forms of these relations indicated in Eq. 14-3. Schamberg [77] has obtained correction terms to the order $\bar{l}^2$ for similar equations for monatomic gases, using the Burnett distribution function. These results are reproduced by Grad's boundary conditions upon iteration in the same manner as the Burnett equations were shown to be obtainable from the Thirteen Moment equations.

Special mention should be made of the term

$$\frac{3}{4} \frac{\mu}{\rho T} \left(\frac{\partial T}{\partial x}\right)_0$$

in Eq. 14-3, according to which a temperature gradient along a surface induces a flow in the direction of increasing temperature. The currents set up by such an induced flow are called the "thermal creep" [61, p. 333]. This phenomenon leads to a pressure gradient along a tube which has a temperature gradient, even when there is no flow, provided that the average pressure is low enough. This can become of some importance in connection with wind tunnel pressure lines and is discussed in I,I.

CHAPTER 4. EXPERIMENTAL RESULTS IN SLIP FLOW AND TRANSITION REGIMES

H,15. Introduction. In the following sections a brief review is presented of the experimental data which are available in the slip and transition regimes for various geometries of aerodynamic interest. A similar review of results for internal flow geometries which are primarily of vacuum engineering interest is given in I,I.

There are also quite a few theoretical analyses available, mostly based upon simplified forms of the Navier-Stokes equations, e.g. the boundary layer equations or the Oseen equations, which include slip and temperature jump boundary condition effects. Often it is assumed that the flow is incompressible, but the results are still presented in terms of the Mach number as well as the Reynolds number. Here the Mach number arises by virtue of the boundary conditions of Eq. 14-3 and the relation $\bar{l}/L \sim M/Re$, so that M is thus not an indication of compressibility at all, but rather of the effect of slip or temperature jump. These analyses are included but should be used with this limitation in mind.

Some of the recent hypersonic experiments have approached the range of slip flow effects, both because of the high Mach number and because of the comparatively low test section densities often used in the hypersonic wind tunnels. It is felt that these experiments are more properly discussed in another place and the present treatment considers only the Mach number range, $0 < M < 6$. For this range the presently known experimental results, particularly for supersonic flow, are due mostly to two laboratories, the Ames Laboratory of the National Advisory Committee for Aeronautics and the Low Density Laboratory of the University of California at Berkeley.

H,16. Couette Flow. The transport of momentum by a rarefied gas between parallel plates or concentric cylinders in relative tangential motion was first studied by Millikan and Van Dyke [*25,83*] for the case of very low speeds. Using the Navier-Stokes equations for the parallel Couette flow, and the slip velocity boundary condition of Eq. 14-3, the drag force per unit area of plate is

$$\tau_w = \tfrac{1}{2}\rho U^2 \frac{2}{Re\left(1 + 2\frac{2 - \sigma}{\sigma}\frac{\bar{l}}{h}\right)} \tag{16-1}$$

U is the relative velocity, h the distance between the plates, $Re = Uh\rho/\mu$, and where it has been assumed that the viscosity μ and the density ρ are constant. This result may be put in the form

$$C_D M = \frac{2}{\dfrac{Re}{M}\left(1 + 2.51\sqrt{\gamma}\dfrac{2 - \sigma}{\sigma}\dfrac{M}{Re}\right)} \tag{16-2}$$

by using Eq. 1-4, where C_D is the appropriate drag coefficient.

For concentric rotating cylinders with very small annular gap, the result is very closely the same. Millikan and Van Dyke used these results to determine σ by measuring the torque transmitted by a rarefied gas across the annular gap between concentric rotating cylinders (see Table H,4b). The calculation has been extended to include the effects of the Burnett terms and of the "second order" slip boundary condition terms by Schamberg [77], and the additional effect of variable viscosity with temperature by Lin and Street [84]. These results are given in the form of power series in l, however, and it has been difficult to determine the number of terms in the series that are consistent with the assumptions made. In view of Wang-Chang and Uhlenbeck's more recent result [47] for the heat transport between parallel plates, the existence of a solution of the fundamental Maxwell-Boltzmann equation in the assumed power series form now seems rather unlikely. A solution of the Thirteen Moment

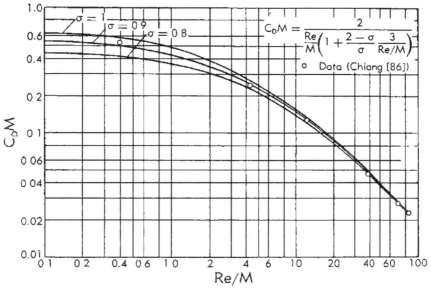

Fig. H,16. Drag coefficient for Couette flow.

equations has also been obtained by Rose [85]. However, to investigate possible deviations from Eq. 16-2, and also to determine how independent of speed the parameter σ is, several investigators have repeated Millikan's experiments recently, at higher speeds. Chiang [86], using air, covered the entire slip and transition range. No change in σ was detectable even for peripheral velocities approaching the sonic range. His results are indicated in Fig. H,16, where the remarkable extent of the validity of Eq. 16-2 is apparent. Similar experimental results have also been obtained by Beams, et al. [87] and Kuhlthau [88].

Incidentally, the limiting value of Eq. 16-2 for $M/Re \to \infty$, is

$$C_D M \to \frac{0.798}{\sqrt{\gamma}\,\frac{2-\sigma}{\sigma}} \tag{16-3}$$

which is the correct value for free molecule flow. This is of course only a coincidence, by which the proper solution of the Boltzmann equation in the low density portion of the transition regime [89] has the same limiting value as Eq. 16-3, as $\bar{l} \to \infty$. The stability of the laminar slip flow between concentric rotating cylinders has been investigated theoretically [90] and experimentally by Chiang [86], with the result that slip has the effect of rendering the flow slightly more stable.

Heat transfer between parallel plates in slip flow may be analyzed in the same way as momentum transfer. If the plates are at different temperatures, T_1 and T_2, and also in relative motion, then the heat flow q per unit area arises from two sources, the viscous dissipation in the flow and the heat conduction due to the temperature difference. From the Navier-Stokes equation and the temperature jump boundary condition, one obtains

$$\frac{hq}{\mu \bar{U}^2} = \frac{1}{\left(1 + 2\dfrac{2 - \sigma}{\sigma}\dfrac{\bar{l}}{h}\right)^2}$$
$$+ \frac{T_1 - T_2}{(\gamma - 1)PrM^2T_0\left(1 + \dfrac{2 - \alpha}{\alpha}\dfrac{4\gamma}{\gamma + 1}\dfrac{\bar{l}}{Prh}\right)} \qquad (16\text{-}4)$$

where T_0 is the average temperature and Pr is the Prandtl number. It is observed that the temperature jump boundary condition reduces the heat transport. The Mach number in all these expressions is a measure of the slip and temperature jump and not of the compressibility.

H,17. Spheres. Slip flow over a sphere at very low Reynolds and Mach numbers has been investigated by Basset [91, p. 270]. Using the Navier-Stokes equations for incompressible flow and the Stokes "slow-flow" simplification of ignoring the inertia terms in comparison to viscous and pressure terms, the effects of slip were incorporated in a boundary condition of the form

$$\tau_t = \beta u_t \qquad (17\text{-}1)$$

where u_t is the tangential slip velocity, τ_t the tangential shear stress, and β the slip coefficient. In terms of Eq. 14-3 one has

$$\beta = \frac{\mu}{\dfrac{2 - \sigma}{\sigma}\bar{l}} \qquad (17\text{-}2)$$

Basset's result for the drag coefficient C_D for a sphere of radius a in a stream of velocity U, density ρ and viscosity μ can be written in the form

$$C_D = \frac{12}{Re}\left(\frac{1 + 2\dfrac{2 - \sigma}{\sigma}\dfrac{\bar{l}}{a}}{1 + 3\dfrac{2 - \sigma}{\sigma}\dfrac{\bar{l}}{a}}\right) \qquad (17\text{-}3)$$

where Re is the Reynolds number $Re = U\rho a/\mu$. A solution of the Thirteen Moment equations for the flow past a sphere, including the effects of compressibility and heat flow, has been obtained by Goldberg [92] which leads to a drag coefficient in the form

$$C_D = \frac{12}{Re}\left[\frac{\left(1 + \frac{15}{2}\frac{2 - \alpha}{\alpha}\frac{\bar{l}}{a}\right)\left(1 + 2\frac{2 - \sigma}{\sigma}\frac{\bar{l}}{a}\right) + \frac{6}{\pi}\frac{\bar{l}^2}{a^2}}{\left(1 + \frac{15}{2}\frac{2 - \alpha}{\alpha}\frac{\bar{l}}{a}\right)\left(1 + 3\frac{2 - \sigma}{\sigma}\frac{\bar{l}}{a}\right) + \frac{9}{5\pi}\left(4 + 9\frac{2 - \sigma}{\sigma}\frac{\bar{l}}{a}\right)\frac{\bar{l}^2}{a^2}}\right]$$

$$(17\text{-}4)$$

As with Couette flow, Eq. 16-3, this result has the same limiting value for $\bar{l}/a \to \infty$ as the free molecule flow result, Eq. 8-6, for $S \to 0$.

Millikan [93] used an approximation to Basset's result in the form

$$C_D = \frac{12}{Re}\frac{1}{1 + \text{const} \cdot \dfrac{M}{Re}} \qquad (17\text{-}5)$$

valid for small M/Re, and an approximation due to Epstein [12] for the free molecule range (i.e. Eq. 8-6 with $S \to 0$) in the form

$$C_D \sim \frac{\text{const}}{M} \qquad (17\text{-}6)$$

Combining Eq. 17-5 and 17-6, Millikan suggested an empirical equation of the form

$$C_D = \frac{12}{Re\left[1 + \dfrac{l}{a}\left(A + Be^{-Cl/a}\right)\right]} \qquad (17\text{-}7)$$

valid over the entire range of M/Re, but restricted to small values of both M and Re. For liquid oil drops in air, Millikan obtained the empirical values of $A = 1.22$, $B = 0.41$, and $C = 8.75$. Eq. 17-7 is the solid curve in Fig. H,17a.

At higher Mach numbers the foregoing analysis breaks down and the situation becomes very complicated due to compressibility effects, including a detached shock wave at supersonic velocities. Experimental values for C_D have been obtained by Kane [94] over the range $2.05 < M < 2.81$; $15 < Re < 768$. No Mach number effect was detected over this range and Kane proposed an empirical formula for C_D in the form

$$C_D = \left(0.97 + \frac{1.32}{\sqrt{Re_1}}\right)\left(1 + \frac{1.0}{\sqrt{Re_1}}\right)^2 \qquad (17\text{-}8)$$

where Re_1 is the Reynolds number based on the flow conditions behind the normal shock and the sphere diameter. This formula is the dashed curve in Fig. H,17a, which also indicates the experimental points for the supersonic experiments. The difference between the dashed and the solid curves is presumably due to the difference in the Mach number. At high Reynolds numbers the skin friction component of the drag is relatively less important than that of the wave drag and the value of the drag coefficient is higher at high Mach numbers. At low Reynolds numbers, however, the skin friction component is of greater relative importance and at the same Reynolds numbers is less at high Mach numbers

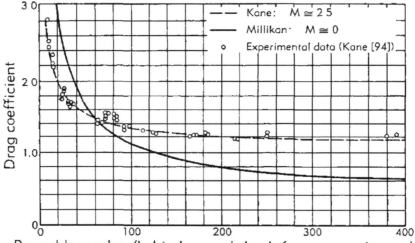

Fig. H,17a.   Sphere drag coefficient.

due to the correspondingly increased effect of slip. The drag coefficient curves for two different Mach numbers thus cross each other, the higher Mach number curve lying below the low Mach number at low Reynolds number and above it at high Reynolds number.

Heat transfer to a sphere in slip flow has been investigated experimentally for supersonic flow by Drake and Backer [95] and Eberley [96], and for subsonic flow by Kavanau [97]. A theoretical estimate of the effect of the temperature jump at the surface on the over-all heat transfer was given in [95] for incompressible flow, using a modified form of Oseen's equation. A somewhat simpler semiempirical analysis for the over-all heat transfer has been suggested by Kavanau, who argued from Eq. 14-3 that the heat transfer with a temperature jump for a sphere of radius a should be the same as the heat transfer without a temperature jump but for a sphere of radius $(a - \text{const } \bar{l})$. Defining the Nusselt number Nu by

$$Nu = \frac{Q}{\pi D(k_{\text{th}})_r(T_w - \dot{T}_r)} \tag{17-9}$$

in the usual notation, with Re based on the diameter, Kavanau obtained by this assumption

$$Nu = \frac{Nu^{(0)}}{1 + 3.42 \dfrac{M}{RePr} Nu^{(0)}} \tag{17-10}$$

where the constant 3.42 has been determined empirically. The quantity $Nu^{(0)}$ is the value the Nusselt number would have if there were no temperature jump, an expression for which has been given by Drake and Backer [95]

$$Nu^{(0)} = 2 + \frac{2}{\pi^2} \int_0^\infty \frac{(1 - e^{-\pi V'})(1 + V^4)^{-1}dV}{J_1^2\left(V\sqrt{\dfrac{2RePrU_{av}}{U}}\right) + Y_1^2\left(V\sqrt{\dfrac{2RePrU_{av}}{U}}\right)} \tag{17-11}$$

Here U_{av} is the "average" velocity over the sphere, which has been taken to be $0.77U$. This equation is the solid curve of Fig. H,17b and H,17c. The supersonic data at $M \sim 2.5$ and $M \sim 3.2$ are shown in Fig. H,17b

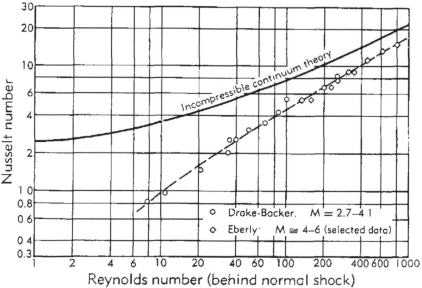

Fig. H,17b. Convective heat transfer coefficient for spheres in supersonic flow.

The subsonic data are shown in Fig. H,17c, together with the corresponding theoretical slip flow and free molecule flow results. The decrease in heat transfer, due primarily to the effective thermal contact resistance associated with the temperature jump, is apparent for both cases.

A somewhat more surprising result than this decrease in the heat

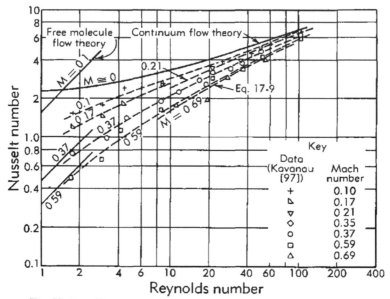

Fig. H,17c. Heat transfer coefficients for spheres in subsonic flow.

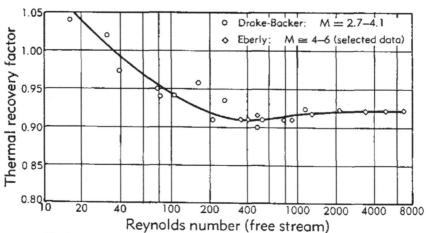

Fig. H,17d. Thermal recovery factor for spheres in supersonic flow.

transfer rate is the trend to values greater than unity of the thermal recovery factor r defined by

$$r = \frac{T_r - T_s}{T_s^0 - T_s} \tag{17-12}$$

where T_r, T_s and T_s^0 are the recovery or adiabatic sphere temperature, the local stream gas temperature, and the local stagnation temperature respectively. This increase is seen in Fig. H,17d. It has already been noted that the recovery factor in free molecule flow for any convex body is greater than unity. It is apparent from the experimental values indicated in Fig. H,17d that this trend starts in the slip flow regime. An analysis has been made [98] of this phenomenon by integrating the energy equation along the forward stagnation stream line, utilizing Homann's velocity distribution [99]. For an adiabatic sphere this leads to an expression for the recovery factor for the stagnation point of the form

$$r \approx 1 + \frac{f(Re)}{Re}, \qquad f(Re) > 0 \tag{17-13}$$

where $f(Re)$ is a slowly varying function. This result, together with the experimental data indicated in Fig. H,17d, suggests that the viscous layer surrounding the sphere is so thick in this Reynolds number range that heat is generated by dissipation more rapidly than it can be conducted away; this is of course quite different from the laminar boundary layer case where the recovery factor is approximately equal to $\sqrt{Pr}$. The result is somewhat similar to the increase in the impact pressure (IX,B,1,) which is also observed at low Reynolds numbers. The experimental accuracy of the subsonic measurements was not sufficient to determine the recovery factor satisfactorily.

H,18. Cylinders. The incompressible flow at low Reynolds number past a right circular cylinder with the axis perpendicular to the free stream velocity was investigated theoretically for the no-slip case by Lamb [100] using the Oseen equations, which partially account for the inertia terms. Tsien [4] extended Lamb's results to the case of slip and determined a drag coefficient in the form

$$C_D = \frac{4\pi}{Re \left[\ln \dfrac{4}{Re} - 1.28 + 1.26 \sqrt{\gamma} \, \dfrac{2 - \sigma}{\sigma} \, \dfrac{M}{Re} \right]} \tag{18-1}$$

where Re is the Reynolds number based on the radius a of the cylinder and C_D is the drag per unit length divided by $a\rho U^2$. There is as yet no experimental data on cylinder drag in the slip flow regime, the results of Finn [101] being just barely in the continuum range, while those of Stalder, Goodwin, and Creager [15] are in free molecule flow.

The heat transfer to cylinders in the slip flow and transition regime has been determined experimentally by Stalder, Goodwin, and Creager [16], Kovásznay and Tormarck [102], and Laufer and McClellan [103]. Their results cover the range $2.0 < M < 3.3$, $0.28 < Re < 203$, and are presented in Fig. H,18a. The Knudsen and Reynolds numbers are based on free stream conditions and the cylinder diameter. The thermal recovery factor, as seen in Fig. H,18b, exhibits the same sort of increase at

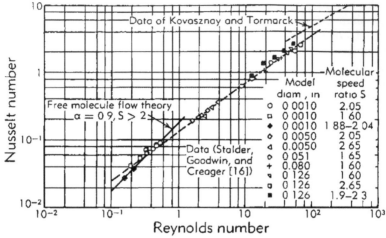

Fig. H,18a. Convective heat transfer coefficients for transverse cylinders in supersonic flow. (Courtesy National Advisory Committee for Aeronautics.)

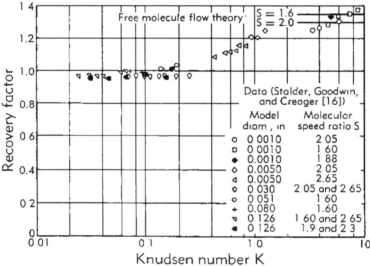

Fig. H,18b. Thermal recovery factor for transverse cylinders in supersonic flow. (Courtesy National Advisory Committee for Aeronautics.)

low densities as that already noted for spheres. A theoretical analysis of the effect of the temperature jump boundary condition on the heat transfer rate for cylinders has been presented by Sauer and Drake [104]. An analysis of heat transfer to cylinders in the Oseen range, neglecting the effect of temperature jump, and thus outside the slip flow regime has also been given by Cole and Roshko [11].

⟨ 44 ⟩

H,19. Flat Plates. Slip flow past a flat plate at zero angle of attack has been studied quite extensively. It was hoped to isolate skin friction phenomena by careful consideration of this geometry. A number of theoretical analyses [7,105,106,107,108,109] have been presented for determining the effect of slip on the boundary layer flow for an isothermal, semi-infinite flat plate. The stream function ψ for the flow, which satisfies the usual boundary layer equations of motion but which obeys the slip boundary condition of Eq. 14-3, can be obtained readily for small l by perturbation series methods in the form,

$$\psi = \psi_0 + \frac{2-\sigma}{\sigma} l \frac{\partial \psi_0}{\partial y} + \cdots \tag{19-1}$$

where y is the normal direction, and where ψ_0 is the stream function for the case of no slip. The local skin friction coefficient c_f is given by

$$c_f = \frac{2\mu}{\rho l'^2} \left(\frac{\partial^2 \psi_0}{\partial y^2} + \frac{2-\sigma}{\sigma} l \frac{\partial^3 \psi_0}{\partial y^3} + \cdots \right)_{y=0} \tag{19-2}$$

However, since

$$\left(\frac{\partial^3 \psi_0}{\partial y^3} \right)_0 \sim \left(\frac{dp}{dx} \right)_\infty = 0$$

for the flat plate boundary layer solution *without pressure gradient*, there is no effect, to order l, of slip on the local skin friction. To order l^2, moreover, the boundary layer equations are no longer valid due, for example, to interaction effects between the viscous layer and the inviscid outer flow [53,54,55]. These interaction effects arise because the boundary layer displaces the inviscid flow outward so that the resulting "free stream" pressure has a negative gradient along the surface; this, in turn, affects the boundary layer so as to increase the skin friction.

A reliable method for extending boundary layer theory to lower Reynolds numbers is not yet available [109,110,111,112,113,114,115,116]. There is thus no adequate theory at present for predicting the effect of slip on the local skin friction of a flat plate at zero angle of attack, the basic difficulty stemming from the fact that the pressure gradient for the ideal flow, and therefore the slip correction term of order l, vanishes. The flat plate thus seems to be an exceptional case. The incompressible boundary layer flow over a wedge of semi-vertex angle θ satisfies the Falkner-Skan equation [117].

$$\psi = \sqrt{u_e \nu x}\, f(\eta), \quad \eta = \sqrt{u_e/\nu x} \cdot y$$
$$mf'^2 - \tfrac{1}{2}(m+1)ff'' = m + f''' \tag{19-3}$$

where the velocity at the edge of the boundary layer is given by $u_e = cx^m$ with $m = \theta/(\pi - \theta)$. This can easily be solved for the case of slip in the form of the perturbation series indicated in Eq. 19-1 to yield an expression for the local skin friction coefficient

$$c_f = c_{f,0} \left[1 - \frac{2-\sigma}{\sigma} \frac{\theta}{\pi - \theta} \frac{\sqrt{Re}}{f''(0)} \frac{l}{x} + \cdots \right] \qquad (19\text{-}4)$$

where $c_{f,0}$ is the coefficient for the no-slip case, and $Re = u_* x/\nu_1$, x being the distance from the vertex. The quantity $f''(0)$ depends on θ, and can be found in [117]. There is thus a nonzero correction term of order l except for the flat plate case of $\theta = 0$. The case $\theta = \pi/2$ is closely related to the stagnation point flow case considered in [105]. Eq. 19-4 may also be put in the form

$$c_f = c_{f,0} \left[1 - \frac{2-\sigma}{\sigma} \frac{\theta}{\pi - \theta} \frac{1.26 \sqrt{\gamma}}{f''(0)} \frac{M}{\sqrt{Re}} + \cdots \right] \qquad (19\text{-}5)$$

by the use of Eq. 1-4; here M is an indication of slip, not of compressibility. It is observed that slip has the expected effect of reducing the local skin friction coefficient.

Experimental values of skin friction in the slip flow range have been determined in both supersonic and subsonic flow [109]. The total drag coefficient based on total drag and the area of one side, for a series of finite flat plates at zero angle of attack, was determined. The increase in the skin friction due to the trailing edge, which has been determined theoretically by Kuo [112] for incompressible no-slip flow, was thus necessarily present. This increase is probably of secondary importance in the supersonic case but explains the increase in the total drag coefficient down to $Re = 15$ observed by Janour [118] in incompressible flow. The experimental data for the slip flow case covered the range $3 < Re < 500$, for $M \cong 0.2$ and 0.6, and $34 < Re < 2020$ for $2.5 < M < 3.8$. This data is presented in Fig. H,19, together with a curve representing the data of Janour (and the theory of Kuo) covering the same Reynolds number range, but for incompressible no-slip flow. It is observed that the measured skin friction for the subsonic case is generally lower than the corresponding continuum results of Janour, in general agreement with Eq. 19-5. The supersonic data, on the other hand, are generally higher than the continuum values. For $Re \cong 1000$ the skin friction increases with increasing Mach number, in agreement with the interaction effect. For $Re \cong 50$, on the other hand, the skin friction decreases with increasing Mach number, in agreement with the slip effect. Skin friction in the slip flow regime is thus apparently affected by two interrelated phenomena of different sign, namely, the interaction of the thick boundary layer with the inviscid flow, and the slip at the surface. The former is more important at the higher densities, gradually giving way to the latter for densities extending into the transition regime.

The question of the degree of slip at the leading edge is of considerable interest, particularly at Mach numbers in the hypersonic range. Using a very small "free molecule flow" equilibrium temperature probe, Laurmann [119] has made a qualitative survey of the leading edge region, of the order of a hundred mean free paths square, at $M \cong 2$ and $M \cong 4$. His results indicate that the "shock wave" is as thick as the viscous

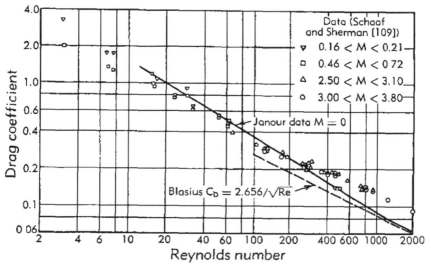

Fig. H,19. Skin friction in slip flow.

layer in this region, that there is a considerable region in which the shock wave coalesces with the viscous layer, and that there appears to be considerable slip on the plate at the higher Mach number.

Heat transfer for a flat plate with a temperature jump boundary condition has been investigated theoretically by Drake and Kane [120] and by Martino [121].

H,20. Cones. For an inviscid supersonic flow, the surface pressure on a cone at zero angle of attack is constant, provided that the Mach number is high enough to insure an attached shock wave at the cone vertex. In the slip and transition regimes, the boundary layer which forms on the cone surface is in general comparatively thick. This displaces the inviscid flow outwards and leads to a pressure distribution which rises rapidly toward the cone vertex. This induced pressure gradient interacts with the boundary layer to increase the skin friction, just as in the flat plate case. Estimating the boundary layer displacement thickness by no-slip flat plate theory and the Mangler transformation to conical flow [122], and using linearized supersonic flow theory to calculate the inviscid flow past a body consisting of the cone plus the boundary layer displacement thickness, Talbot [*123*] has obtained a result for the induced pressure in the form

$$\frac{p}{p_{\text{ideal}}} = 1 + \frac{f(M_1, \theta)}{\sqrt{Re_1}} \tag{20-1}$$

where θ is the cone semi-vertex angle, M_1 the Mach number, and Re_1 the Reynolds number based on conditions on the cone surface in inviscid

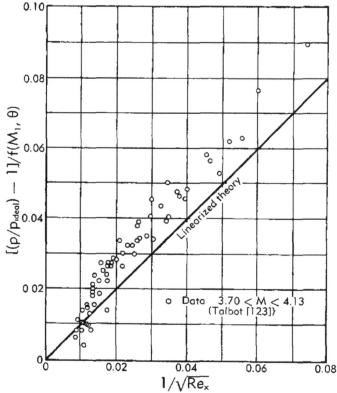

Fig. H,20a. Surface pressure distribution for cones with 5° semi-vertex angle.

flow and the slant distance x from the vertex. The quantity $f(M_1, \theta)$ is given by

$$f(M_1, \theta) = \frac{\gamma M_1^2 \left(3\theta \ln \dfrac{1}{\theta} - 4\theta \right) (1 + 0.277 M_1^2)}{4[1 + \theta^2(1 + 2 \ln \theta)]} \tag{20-2}$$

Talbot has measured the surface pressure on slender (5° half angle) cones in the range $3.70 < M < 4.13$, $185 < Re < 12,000$. His results are indicated in Fig. H,20a, along with the theoretical result correspond-

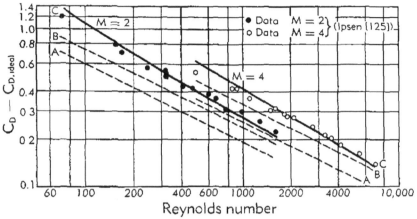

Fig. H,20b. Drag coefficients for cones with 15° semi-vertex angles.

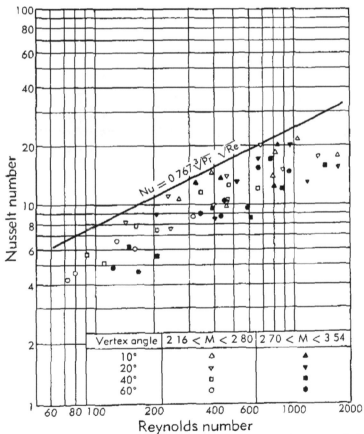

Fig. H,20c. Convective heat transfer coefficients for cones in supersonic flow

ing to Eq. 20-1, confirming the general trend of the prediction. M and Re are based on free stream conditions.

The most complete theoretical estimate of the skin friction on cones in this range has been made by Probstein and Elliott [*124*], considering the interaction effect mentioned above and also the effect of transverse curvature on the boundary layer flow. Ipsen [*125*] has measured the

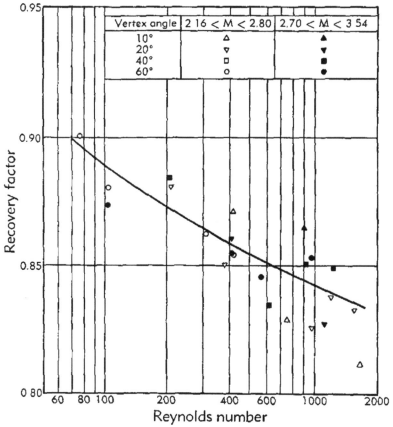

Fig. H,20d.   Thermal recovery factors for cones in supersonic flow.

total drag on 15° half angle cones over the range $150 < Re < 1500$ at $M = 2$ and $1000 < Re < 7000$ at $M = 4$. His results are presented in Fig. H,20b. Here M and Re are based on the free stream conditions and the cone slant length. The quantity C_D is the drag coefficient based on cone surface area, while $C_{D, \text{ideal}}$ is the corresponding inviscid drag coefficient at the same Mach number. Curves A, B, and C are the calculated values of $C_D - C_{D, \text{ideal}}$ accounting for, respectively, the simple boundary layer skin friction, this plus the induced pressure, and finally this plus the transverse curvature correction to the skin friction. Again, none of these theories takes account of possible slip effects. The slight'drop-off of the data for the lowest values of *Re* at each Mach number might possibly be due to such slip effects, although neither the experiments nor the theories are of sufficient accuracy to warrant such a conclusion.

The convection heat transfer characteristics of cones in the slip flow range have been determined by Drake and Maslach [126] over the range $78 < Re < 3270$, $2.16 < M < 3.54$ where M and Re are based on the free stream conditions and the cone slant length. The corresponding variation of the Nusselt number and the recovery factor with the Mach and Reynolds numbers are presented in Fig. H,20c and H,20d. The decrease in the heat transfer rate as well as the increase in the recovery factor with decreasing density are again observed.

H,21. Base Pressure. The base pressure p_b on a cone-cylinder configuration at zero angle of attack has been determined in the slip flow and transition regimes by Kavanau [127,128]. His results, together with a curve indicating the continuum results for geometrically similar models

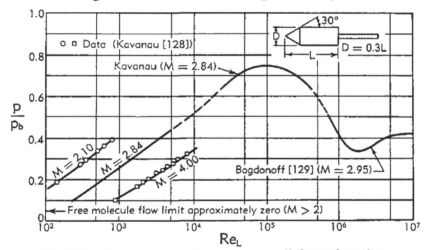

Fig. H,21. Base pressure coefficient for cone-cylinder configurations.

obtained by Bogdonoff [129], are given in Fig. H,21. The Reynolds number is based on the free stream and the model length, while p_∞ is the free stream static pressure. The peak in the base pressure curve near $Re = 10^6$ is interpreted as the point at which the transition point coincides with the critical point in the wake. An increase in Reynolds number introduces some turbulent mixing in the wake region, thus increasing the over-all mixing and hence decreasing the base pressure. On the other hand, a decrease in Reynolds number increases the laminar mixing and thus also decreases the base pressure. For the measurements in the lower Reynolds number range, Kavanau found that the pressure on the base varied considerably with radial position, being as much as four times as high at the center as at the periphery. The base pressures indicated in Fig. H,21 are area averages.

H,22. Cited References.

1. Maxwell, J. C. *The Scientific Papers of James Clerk Maxwell*, Vol. 2. Cambridge Univ. Press, 1890.
2. Kennard, E. H. *Kinetic Theory of Gases*. McGraw-Hill, 1938.
3. Loeb, L. B. *Kinetic Theory of Gases*, 2nd ed. McGraw-Hill, 1934.
4. Tsien, H. S. *J. Aeronaut. Sci. 13*, 653 (1946).
5. Tsien, H. S. *J. Aeronaut. Sci. 15*, 577 (1948).
6. Roberts, H. E *Aeronaut. Eng. Rev. 8*, 19 (1949).
7. Donaldson, C. duP. An approximate method for estimating the incompressible laminar boundary layer characteristics on a flat plate in slipping flow *NACA Research Mem. L9C02*, 1949.
8. Siegel, K. M. *J. Aeronaut Sci. 17*, 191 (1950).
9. Van de Hulst, H. C. *The Atmospheres of the Earth and Planets*. Univ. Chicago Press, 1952.
10. Warfield, C. N. Tentative tables for the properties of the upper atmosphere *NACA Tech. Note 1200*, 1947.
11. Cole, J., and Roshko, A. Heat transfer from wires at Reynolds numbers in the Oseen range. *Heat Transfer and Fluid Mech. Inst.*, 1954.
12. Epstein, P. S. *Phys. Rev. 23*, 710 (1924).
13. Zahm, A. F. *J. Franklin Inst. 217*, 153 (1934).
14. Grimminger, G., Williams, E P., and Young, G. B. W *J. Aeronaut. Sci. 17*, 241 (1950).
15. Stalder, J. R., Goodwin, G , and Creager, M. O. A comparison of theory and experiment for high speed free molecule flow. *NACA Tech Note 2244*, 1950.
16. Stalder, J. R., Goodwin, G., and Creager, M. O. Heat transfer to bodies in a high speed rarefied gas stream. *NACA Tech. Note 2438*, 1951.
17. Sänger, E. Gaskinetik Sehr hoher Fluggeschwindigkeiten. *Deut. Luftfahrt-forschung, Bericht 972*, 1938. Transl. as: The gas kinetics of very high altitude flight. *NACA Tech. Mem. 1270*, 1950.
18. Smoluchowski, M. V. *Wied. Ann. 64*, 101 (1898).
19. Knudsen, M. *Ann. Phys. 34*, 593 (1911).
20. *Hand und Jahrbuch der Chemischen Physik*, Vol. 3, Pt. II, Sec. IV. Akad. Verlagsgesellschaft, Leipzig, 1949.
21. Wiedmann, M. L. *Trans. Am. Soc. Mech. Engrs. 68*, 57 (1946).
22. Devienne, P. F. M. Condensation et adsorption des molécules sur une surface en atmosphère raréfiée. *Mém. sci. phys. 52*, 1952.
23. Devienne, P. F. M. Conduction thermique dans les gaz raréfiés coefficient d'accommodation. *Mém. sci. phys. 56*, 1953.
24. Hurlbut, F. C. *An Experimental Molecular Beam Investigation of the Scattering of Molecules from Surfaces*. Ph.D. Thesis, Univ. Calif., 1954.
25 Millikan, R. A. *Phys. Rev. 21*, 217 (1923).
26. Sänger, E., and Bredt, J. Uber einen Raketenantreib fur Fernbomber. *Deut. Luftfahrtforschung, Untersuch. u. Mitt. 3538*, 141 (1944)
27. Luntz, M. *Recherche aéronaut. Paris 7*, 17 (1949).
28. Stalder, J. R., and Jukoff, D. Heat transfer to bodies travelling at high speeds in the upper atmosphere. *NACA Rept. 944*, 1949.
29. Oppenheim, A. K. *J. Aeronaut. Sci*, 49 (1953).
30. Sauer, F. M. *J. Aeronaut. Sci. 18*, 353 (1951).
31. Hartree, D. R. *Mem. and Proc. Manchester Lit. and Phil Soc. 80*, 85 (1935)
32. Heineman, M. *Commun. on Pure and Appl. Math. 1*, 259 (1948).
33. Snow, R. M. Aerodynamics in a highly attenuated atmosphere. *Appl. Phys. Lab., Johns Hopkins Univ. Rept. CM336*, 1946.

34. Ashley, H.   *J. Aeronaut. Sci. 16,* 95 (1949).
35. Stalder, J. R., and Zurick, V. J.   Theoretical aerodynamic characteristics of bodies in a free molecule flow field. *NACA Tech. Note 2423,* 1951.
36. Popov, S. G.   *Vestnik Moskovskozo Universiteta 3,* 25 (1948). (*Am. Math. Soc. Transl. 36.*)
37. Garfunkel, I. M.   Generalizations of applications of free molecule flow. *Univ. Mich Rept. EMB58.*
38. Bell, S., and Schaaf, S. A.   *J. Am. Rocket Soc. 23,* 314 (1953).
39. Bell, S., and Schaaf, S. A.   *Jet Propulsion 25,* 168 (1955)
40. Sherman, F. S.   A low density wind tunnel study of shock wave structure and relaxation phenomena in gases. *NACA Tech Note 3298,* 1955.
41. Burgers, J. M.   *Proc. Kon. Vlaamse Acad. voor Wet.,* 573 (1947).
42 Thomas, R. N., and Whipple, F. L.   *J. Aeronaut. Sci. 18,* 636 (1951).
43. Sanger, E.   *Schweiz. Arch. 16,* 43 (1950).
44. Jaffé, G.   *Ann. Physik 6,* 195 (1930).
45. Keller, J.   *Commun. on Pure and Appl. Math. 1,* 275 (1948).
46. Wang-Chang, C. S.   Transport phenomena in very dilute gases. *Nav. Ord. Research and Develop. Rept. 7924-UMHSF,* 1950.
47. Wang-Chang, C. S., and Uhlenbeck, G. E.   The heat transport between two parallel plates as functions of the Knudsen number. *Univ. Mich. Eng. Research Inst. Rept. M999,* 1953.
48. Szymanski, Z.   *Arch. Mech. Stosowanej 8,* 1956.
49. Lunc, M.   *Technika Lotnicza 5,* 39 (1950).
50. Lunc, M., and Lubonski, J.   *Arch. Mech. Stosowanej 4,* 597 (1956).
51. Kryzwoblocki, M. Z., and Shinosaki, G.   *Acta Phys. Austriaca 10,* 34–53 (1956).
52. Bowyer, J. M.   *Drag on a Rotating Cylinder Apparatus near Conditions of Free Molecule Flow. Ph.D. Thesis,* Univ. Calif., 1956.
53. Lees, L., and Probstein, R. F.   Hypersonic viscous flow over a flat plate. *Princeton Univ. Aeronaut. Eng. Dept. Rept. 195,* 1952
54. Lees, L.   *J. Aeronaut. Sci. 20,* 143–145 (1953).
55. Li, T. Y., and Nagamatsu, H. T.   *J. Aeronaut. Sci 20,* 345–351 (1953).
56. Grad, H.   *Commun. on Pure and Appl. Math. 2,* 331 (1949).
57. Koga, T.   *J. Chem. Phys. 22,* 1633–1646 (1954).
58. Enskog, D.   *Arkiv Mat, Astron. Fysik Stockholm 21,* 13 (1928).
59. Burnett, D.   *Proc. London Math. Soc. 40,* 382 (1935).
60. Chapman, S., and Cowling, T. G.   *The Mathematical Theory of Non-Uniform Gases.* Cambridge Univ. Press, 1939.
61. *Heat Transfer.* Univ. Mich. Press, 1953.
62. Hirschfelder, J. O., Curtiss, C. F., and Bird, R. B.   *Molecular Theory of Liquids.* Wiley, 1954.
63. Patterson, G. N.   *Molecular Flow of Gases.* Wiley, 1956.
64. Greene, E. F, Cowan, G. R, and Hornig, D. F.   *J. Chem. Phys. 19,* 427 (1951).
65. Greenspan, M.   *J Acoust. Soc. Amer. 26,* 70 (1954).
66. Rosenhead, L., et al.   A discussion on the first and second viscosity of fluids. *Proc. Roy. Soc. London A226,* 1–65 (1954).
67. Wang-Chang, C. S., and Uhlenbeck, G. E.   Transport phenomena in polyatomic gases. *Univ. Mich. Eng. Research Inst. Rept. CM681,* July 1951.
68. Truesdell, C.   *Proc. Natl. Acad. Sci. 34,* 342 (1948).
69. Truesdell, C.   *J. Rat. Mech. and Anal. 1,* 125–300 (1952).
70. Mostov, P. M., Grad, H., and Borowitz, S.   *Phys. Rev. 90,* 376 (1953).
71. Kohler, M.   *Z. Physik 124,* 757 (1948).
72. Andersen, W. H., and Hornig, D. F.   *J. Chem. Phys. 24,* 767–770 (1956).
73. Talbot, L., and Sherman, F. S.   Structure of weak shock waves in a monatomic gas. *NACA Tech. Note.* (In press.)
74. Wang-Chang, C. S., and Uhlenbeck, G. E.   On the behavior of a gas near a wall. *Univ. Mich. Eng. Research Inst. Rept. 2457-2-T,* Aug. 1956.
75. Ikenberry, E., and Truesdell, C.   *J. Rat. Mech. and Anal. 5,* 1–54 (1956).

76. Truesdell, C. *J. Rat. Mech. and Anal. 5*, 55–128 (1956).
77. Schamberg, R. *The Fundamental Differential Equations and the Boundary Conditions for High Speed Slip Flow.* Thesis, Calif. Inst. Technol., 1947.
78. Smoluchowski, M. V. *Ann. Physik 33*, 1559 (1910).
79. Weber, S. *Det. Kgl. Danske Videnskabernes Selskab Math. fysiske Meddeleiser 16*, 9 (1939).
80. Kramers, H. A. *Nuovo Cimento 6, Suppl.*, 1949.
81. Welander, P. *Arkiv för Fysik 7*, 507–553 (1953).
82. Payne, H. *J. Chem. Phys. 21*, 2127–2132 (1953).
83. Van Dyke, K. S. *Phys. Rev. 21*, 250 (1923).
84. Lin, T. C., and Street, R. E. Effect of variable viscosity and thermal conductivity on high-speed slip flow between concentric rotating cylinders. *NACA Tech. Note 2895*, 1953.
85. Rose, M. *Phys. Rev. 91*, 469 (1953).
86. Chiang, S. F. *Drag Forces on Rotating Cylinders at Low Pressures.* Ph. D. Thesis, Univ. Calif., 1952.
87. Beams, J. W., Young, J. L., and Moore, J. W. *J. Appl. Phys. 17*, 886 (1946).
88. Kuhlthau, A. R. *J. Appl. Phys. 20*, 217 (1949).
89. Wang-Chang, C. S., and Uhlenbeck, G. E. Transport phenomena in very dilute gases. *Univ. Mich. Eng. Research Inst. Rept. CM579*, Nov. 1949.
90. Chambré, P. L. *Stability of Viscous Compressible Flow Between Rotating Cylinders. Ph.D. Thesis*, Univ. Calif., 1951.
91 Basset, A. B. *Hydrodynamics.* Deighton, Bell & Co., Cambridge, 1888.
92. Goldberg, R. *The Flow of a Rarefied Perfect Gas Past a Fixed Spherical Obstacle. Ph.D. Thesis*, New York Univ., 1954.
93. Millikan, R. A. *Phys. Rev. 22*, 1 (1923).
94. Kane, E. D. *J. Aeronaut. Sci. 18*, 259 (1951).
95. Drake, R. M., and Backer, G. H. *Trans. Am. Soc. Mech. Engrs. 74*, 1241–1250 (1952).
96. Eberley, D. K. Forced convection heat transfer from spheres to a rarefied gas. *Univ. Calif. Inst. Eng. Research Rept. HE-150-140*, July 1956.
97. Kavanau, L. L. *Trans. Am. Soc. Mech. Engrs. 77*, 617–624 (1955).
98. Mack, S. F., and Schaaf, S. A. Viscous effects on stagnation point temperatures. *Univ. Calif. Rept. HE-150-96*, 1951.
99. Homann, L. *Z. angew. Math. u. Mech. 16*, 153 (1936).
100. Lamb, H. *Z. agnew. Math. u. Mech. 16*, 615 (1936)
101. Finn, R. K. *J. Appl. Phys. 24*, 771 (1953).
102. Kovásznay, L. S. G., and Tormarck, S. I. A. Heat loss of hot wires in supersonic flow. *Johns Hopkins Univ. Aeronaut. Dept. Bumblebee Rept. 127*, Apr. 1950.
103. Laufer, I., and McClellan, R. Equilibrium temperature and heat transfer characteristics of hot wires in supersonic flow. *Guggenheim Aeronaut. Lab., Calif. Inst. Technol. External Rept. 315*, May 1956.
104. Sauer, F. M., and Drake, R. M. *J. Aeronaut. Sci. 20*, 175 (1953).
105. Lin, T. C., and Schaaf, S. A. The effect of slip on flow near a stagnation point and in a boundary layer. *NACA Tech. Note 2568*, 1951.
106. Mirels, H. Estimate of slip effect on compressible laminar boundary layer skin friction. *NACA Tech. Note 2609*, 1952.
107. Maslen, S A. Second approximation to laminar boundary layer flow on flat plate in slip flow. *NACA Tech. Note 2818*, 1952.
108. Nonweiler, T. The laminar boundary layer in slip flow. *College of Aeronautics, Cranfield, Rept. 62*, Nov. 1952.
109. Schaaf, S. A., and Sherman, F. S. *J. Aeronaut. Sci. 21*, 85 (1954).
110. Alden, H. L. *J. Math. and Phys. 47*, 91 (1948).
111. Lewis, J. A., and Carrier, G. F. *Quart. Appl. Math. 7*, 228 (1949).
112. Kuo, Y. H. *J. Math. and Phys. 32*, 83 (1953).
113. Kaplan, S. *J. Appl. Math. and Phys. 5*, 111–136 (1954).
114. Latta, G. E. Investigations of the classical boundary layer associated with

Oseen's equation. *Guggenheim Aeronaut. Lab., Calif. Inst. Technol.* Unpublished manuscript.
115. Lagerstrom, P. A., and Cole, J. D. *J. Rat. Mech. and Anal. 4,* 817–882 (1955).
116. Goldstein, S. Summary interim report—Flow of an incompressible viscous fluid along a semi-infinite flat plate. *Univ. Calif. Inst. Eng. Research Rept. HE-150-144,* Dec. 1956.
117. Falkner, V. M., and Skan, S. W. Some approximate solutions of the boundary layer equations. *Brit. Aeronaut. Research Council Repts. and Mem. 1314,* Apr. 1930.
118. Janour, Z. Resistance of a plate in parallel flow at low Reynolds numbers. *NACA Tech. Mem. 1316,* 1951.
119. Laurmann, J. A. Experimental investigation of the flow about the leading edge of a flat plate. *Univ. Calif. Inst. Eng. Research Rept. HE-150-126,* Oct. 1954.
120. Drake, R. M., and Kane, E. D. A summary of the present status of heat transfer in rarefied gases. *Univ. Calif. Inst. Eng. Research Rept. HE-150-73,* Oct. 1950.
121. Martino, R. L. Heat transfer in slip flow. *Univ. Toronto Inst. Aerophys. Rept. 35,* 1955.
122. Mangler, W. *Z. angew. Math. u. Mech. 28,* 97 (1948).
123. Talbot, L. Viscosity corrections to cone probes in rarefied supersonic flow at a nominal Mach number of 4. *NACA Tech. Note 3219,* 1954.
124. Probstein, R. F., and Elliott, D. The transverse curvature effect in compressible axially-symmetric laminar boundary layer flow. *J. Aeronaut. Sci. 23,* 208–225 (1956).
125. Ipsen, D. C. *Cone Drag in a Rarefied Gas Flow. Ph.D. Thesis,* Univ. Calif., 1953.
126. Drake, R. M., and Maslach, G. J. Heat transfer from right circular cones to a rarefied gas in supersonic flow. *Univ. Calif. Inst. Eng. Research Rept. HE-150-91,* 1952.
127. Kavanau, L. L. *J. Aeronaut. Sci. 21,* 257 (1954).
128. Kavanau, L. L. Base pressure studies in rarefied supersonic flow. *J. Aeronaut. Sci. 23,* 193–208 (1956).
129. Bogdonoff, S. M. *J. Aeronaut. Sci. 19,* 201 (1952).

Milton Keynes UK
Ingram Content Group UK Ltd.
UKHW050121240624
444575UK00005B/335